COLORING
TECHNOLOGY

for Plastics

Ronald M. Harris, Editor

Society of Plastics Engineers

Plastics Design Library

Table of Contents

PREFACE

A lot has happened in the field of Plastics Coloring Technology over the past 4 years. The papers in this volume, which were presented during 7 ANTEC and RETEC symposia in 1995-1998, chronicle many of the advances. As you read through the collection, you will find progress took place on a number of fronts. There is, however, a unifying theme: creating more value.

Color has always been an important component of the plastics "value proposition". In the beginning, putting color into plastics justified its cost by eliminating, in many cases, the need to paint a fabricated part. This is no longer sufficient. Today, color stylists seek to exploit the entire appearance experience to help create a desired product image. No, you can't tell a book by its cover, but a product's package can generate appeal and additional sales. By packaging laundry detergent in a bright fluorescent jug, for example, the manufacturer not only attracts shoppers to the store shelf but conveys a vision of the outcome of using its product – bright, clean clothes. Package design has been an interactive process: stylists demanding more from the technology, and, at the same time, advances in plastics coloring technology stirring creative juices.

Ironically, one way in which our industry has created more value is by making plastic look more like the traditional materials they replaced: glass, wood, stone, and metal. To many consumers "plastic" has stood for "artificial" and "cheap." As discussed in many of the papers in the first section of this collection ("Pigments and Dyes"), the industry has responded with special effects colorants to give plastic the luster and iridescence of pearl, the rich sheen of gold and silver, or natural appearance of wood, stone, or leather. This section also contains a pair of timely articles covering health and safety aspects of plastics colorants.

Being responsive is another way to create value. "Speed to market" is the new catch phrase. Packagers, auto makers, appliance manufacturers, and other users of colored plastic have to respond quickly to wants and needs of their consuming public, each competitor seeking to differentiate its products. You really can have (almost) any color you want when you go to purchase a car. Partnering with responsive suppliers is a critical success factor. I cited above the interaction between product stylists and colorists. That is just one aspect of being responsive. Another is addressing the operational needs of the fabricator. Inventory is a dirty word. Fabricators want their color just in time, and the need for speed makes it imperative to have first run acceptance on color. Adjusting color is not just a waste of money; it limits your ability to respond and puts you and your customer at risk of losing business. The second ("Effective Pigment Incorporation") and third sections ("Testing Colored Products") deal with advances in processing, color measurement, and assaying additives that can help color houses be more responsive. You may find the paper on in-line color measurement of special interest.

Speed is one aspect of being responsive; consistency is another. Testing in-coming raw materials, including pigments, is an avoidable expense. Whether you are formulating color or using formulated color, starting up a new run of your colored product should not have to be an adventure. In addition to the processing papers on pigment incorporation, we have also included a paper in the first section of the book describing a novel pigment manufacturing process designed to give improved lot-to-lot color consistency.

Ten years ago the buzzword was "quality." Claims of superior quality are no longer sufficient to differentiate your product. On the other hand, quality failures can be fatal. Color related defects of any kind potentially mar a product's image. An auto's performance will not depend on the precision of the color match between the polypropylene instrument panel and the vinyl dashboard. It will, however, influence your decision to buy that particular car. Or, if the color fades within the first year you will more than likely tell others – as well as lodge a complaint. The papers in the third ("Testing Colored Products") and fourth sections ("Effects of Colorants on Colored Materials") deal with our enhanced understanding of the effects of colorful additives on plastics and advances in technology to characterize colored materials and assure their performance in the field.

Another way color suppliers bring value to their customers is through applications assistance. If this term is unfamiliar to you, think of it as the proactive side of technical service. Applications assistance can be as mundane as instructions on how to use the product or as forward thinking as helping a customer win new business. Do not underestimate its value. In a maturing industry such as coloring plastics, applications assistance is one way to continue to differentiate your company. This volume offers articles on pigment selection, coloring "do's and don'ts", effects of injection molding parameters, color matching, and the like. Some of the information presented in these articles, such as this author's "A Primer on Colorful Additives", is well known to practitioners. However, the industry is only beginning to appreciate the control the fabricator has on color and appearance of the finished parts through machine set-up conditions. The articles describing the coloring of polymer blends and thin-walled parts also present timely information.

We also bring value to our customers by enhancing their capabilities. Rule number one: customers do not buy what you make, they buy what they need. The entire supply chain can be viewed in terms of suppliers at one level meeting the needs of suppliers at the next. Success here is determined not so much by what you produce as it is by your capability to create solutions to customer needs (read that "problems"). The emerging field of Laser Marking provides an example of creating more value by enhancing capability. Laser technology enables a fabricator to produce short runs of custom marked parts in a cost effective manner. These markings can be functional codes, promotional sayings or symbols, or even decorative labels. The final section of this volume offers a half dozen articles covering principles and applications of this new technology.

What are some other new directions for Plastics Coloring Technology? I believe this collection gives us a hint. We have a long way to go in developing response capabilities. In the next few years expect to see further advances in on-line color and additive measurements, including on-line color adjustment. In the area of special effects, interference colorants have been taken to another level recently by the introduction of flake pigments produced by vacuum deposition techniques. They have made their way into coatings applications. We will soon find them in plastics. Another emerging technology is "smart" color. The two articles in this collection dealing with photochromic dyes give us peek at the future. We will see further advances in all areas of color change technology, including thermo- and piezochromic effects, and in the development of devices that utilize color change.

Hang on. The ride is not over.

Ronald M. Harris, Ph.D.
Ferro Corporation
Cleveland, Ohio
January, 1999

A Primer on Colorful Additives

Ronald M. Harris
M. A. Hanna Color Technical Center, Suwanee, Georgia, USA

INTRODUCTION

The melt coloring of plastics is one of the most functional value added features a resin producer, compounder, or parts fabricator can impart to their products. It not only provides desired appearance properties that help sell the product, but it can also enhance several other properties, such as stability toward UV light. In addition, melt coloring usually eliminates the need for a separate, off-line, painting step. Overall manufacturing costs can thereby be reduced.

Once the color system is incorporated into the plastic matrix, however, it becomes an integral part of the material and may alter the engineering, performance, and processing properties in ways not considered during the design and formulation of the new material. Coloring, frankly, is usually viewed as the end-users problem, and the ball lands in the court of the color formulator. This specialist (who is usually untrained in the finer points of polymer science) is then left with the task of navigating through what often seems to be an obstacle course of known and unpredicted interactions while trying to give the end-user an economical color package that will meet the product's appearance requirements. The task is even more critical in the case of high performance polymer blends and alloys, whose highly valued engineering and performance properties are often sensitive to small compositional changes.

The objective of this Section is to raise the level of awareness that color needs to be part of any total systems approach to material design. We will survey the major classes of colorants suitable for use in high performance polymer blends and alloys and describe some of the potential chemical and physical colorant/material interactions.

THE MAJOR CLASSES OF COLORANTS

The colorants used in plastics fall into two very broad categories: pigments and dyes. Pigments are defined as colorants which do <u>not</u> dissolve in the plastic matrix of interest, whereas dyes are colorants that do go into solution. Pigments, therefore reside as a separate phase. Consequently, there are phase boundaries to consider, and these can be crucial to the end-user.

INORGANIC PIGMENTS

Inorganic pigments are metal salts and oxides which can predictably impart color to a substrate. Most of these pigments have an average particle size of about 0.2-1.0 microns. The manufacturers take great pains to eliminate agglomerates with particle sizes above 5 microns. With few exceptions, inorganic pigments are inexpensive raw materials (see tables which follow). Because of their relatively low color strength they are not always the best value. Some good properties which many inorganic pigments share are:

1   easy to disperse (relatively little work is required to break down the pigment, coat it with the plastic, and distribute it uniformly);
2   good heat and weather resistance;
3   little, if any, reactivity.

There are, of course, exceptions, and we will be quick to point these out. For example, reactivity is a potential problem in any polymer system.

Simple Metal Oxides

Several inexpensive colorants compatible with a wide variety of polymers have a simple chemical structure - they are single metal oxides. For convenience, we have listed the four most commonly used examples in Table 1, along with several properties of interest, such as compliance with food contact applications and cost.

Titanium Dioxide (Titania) Titanium dioxide is the least expensive and most widely used white pigment. Nearly all of the titania used in this country is produced from titanium ore (mostly titanium and iron oxides) by the "chloride process", which goes through a $TiCl_4$ intermediate.

Although TiO_2 is viewed as an inert chemical, the crystal is defective - lacking 1 oxygen atom in about 100,000. This defect results in each titania particle (ca. 0.2-0.3 microns) having a small number of reactive Ti(+) sites on or near the surface. In chloride process titania, some of these sites have bound chloride ions. Ti(+) ions and bound Cl(-) ions can be double trouble for some blends and alloys. Titanium ions can react with a variety of organic substances to form organic titanates that can catalyze all kinds of reactions. Polymer chemists, for example, use titanates to catalyze polymerization of polyesters. In addition, the chloride ion promotes the degradation of polycarbonate, PC. Some grades of titania are therefore unsuitable for use in blends or alloys containing PC.

Table 1. Inorganic pigments

Pigment	Strength	Heat stability	Weather resistance	FDA status	Cost $/lbs
Simple metal oxides					
Titanium dioxide (Pigment White 6)	G	G-E	G-E	yes	1.0
Zinc oxide (Pigment White 4)	P	G-E	G-E	yes	1.0
Iron oxides (Pigment Red 101)	F	G-E	G-E	yes	1.0
Chrome (III) oxide (Pigment Green 17)	F	E	E	yes	2.0
Mixed metal oxides					
Nickel titanate (Pigment Yellow 53)	P	E	E	no	4.0
Nickel chrome titanate (Pigment Brown 24)	P	E	E	no	4.0
Copper chromate (Pigment Black 28)	P	E	E	no	6.0
Cobalt chromite (Pigment Green 26)	F	E	E	no	8.0
Cobalt aluminate (Pigment Blue 28)	F	E	E	yes	15.0
Metal sulfides					
Zinc sulfide (Pigment White 7)	F	G-E	G	yes	1.5
Cadmium sulfide (Pigment Yellow 35,37)	G	E	F	no	12.5
Cadmium sulfoselenide (Pigment Orange 20)	G	E	G	no	17.0
Cadmium sulfoselenide (Pigment Red 108)	G	E	E	no	22.0
Ultramarine pigments					
Ultramarine blue (Pigment Blue 29)	F	E	P-F	yes	2.0
Ultramarine violet (Pigment Violet 15)	P	E	P-F	yes	4.5

To minimize the reactivity of this important white pigment TiO_2 producers process special plastics grade versions in which the reactive sites are masked with alumina, silica, and/or silicone fluid *via* proprietary means.

Both crystal forms of titania, rutile and anatase, have a high Mohs hardness rating and are abrasive. For example, the pigment particles can etch glass fibers and greatly reduce the tensile strength of glass fiber reinforced thermoplastic blends and alloys.

<u>Synthetic Iron Oxide Reds</u> The synthetic iron oxide red pigments are based on Fe_2O_3 chemistry.

These pigments offer a great deal of hiding power (opacity), but are relatively weak and dirty. A wide range of rust colored hues are available. Again, the Fe(+3) acts as a Lewis acid. Even at low concentrations it poses a problem for PVC and is not suitable for blends and alloys containing PVC. At higher concentrations it may adversely affect some PC materials.

Chrome(III) Oxide Chromic oxide is a weak, olive shade green with excellent weathering and heat stability. Because other green pigments with higher color strength and cleaner hues are available, chrome oxide pigments are usually reserved for long-term outdoor applications.

Mixed Metal Oxide Pigments

The mixed metal oxides are a large class of calcined pigments (final step involves a slow heat treatment at about 1000°C) developed originally for ceramics, that have excellent weatherability and heat stability. As a class, they are the least reactive pigments and, for example, have enjoyed wide use in vinyl siding. On the down side, these pigments have low tinting strength and, as a rule, are abrasive toward glass fibers. Properties of some commonly used MMO pigments are listed in Table 1.

Many of the MMO pigments contain significant levels (about 10% by weight) of antimony, which is considered a heavy metal by some regulatory agencies. Antimony free versions are now available from several suppliers, but there is a cost penalty of about 25-50%.

In the past, MMO pigments were found to be hard to disperse and the source of color specks. Recently, however, all of the major suppliers have targeted polyamide fibers as a lucrative new market and have greatly improved the particle size distribution and ease of dispersion. Some combinations of metals and the colors produced are given in Table 1.

Metal Sulfide Pigments

Included among these pigments are zinc sulfide (ZnS) and the cadmium-based pigments. ZnS is a white pigment which has found a niche in coloring glass reinforced resins. The cadmium pigments represent a large class of bright pigments in the yellow-orange-red range of the color spectrum. Some properties of metal sulfide pigments are listed in Table 1.

Zinc Sulfide White ZnS is a white pigment having about one-half the opacifying and tinting strength of titania. Two special features are:

1 its low Mohs hardness recommends it for use in glass reinforced resins
2 it exhibits much lower light absorption in the near UV region of the spectrum than titania. Because of this, it can be used with some bright fluorescent dyes with   minimal quenching of fluorescence.

Zinc(+2), as noted above, is a Lewis acid, and most color formulators avoid the use of ZnS in PC and its blends and alloys.

Cadmium Sulfides and Sulfoselenides Calcined CdS is a bright orange shade yellow with excellent light fastness and heat stability. It does not hold up well in moist atmospheres and is not necessarily suitable for all outdoor applications. By combining Cd and Zn (up to about 12% Zn) bright primrose and lemon yellow hues can be achieved. Replacing some sulfur with selenium shifts the color toward the red region. Total replacement yields CdSe, a deep maroon pigment.

Traditionally, the cadmium-based pigments have been widely used in engineering resins such as the polyamides 6 and 6/6, PC, thermoplastic polyesters and their blends and alloys.

Cadmium, however, is classified as a heavy metal by all regulatory agencies. For example, Cd is one of the four heavy metals cited in the CONEG legislation governing packaging materials. More and more, end-users of engineered materials - including blends and alloys - are specifying heavy metal free formulations, and cadmium-based pigments are slowly, but surely, being replaced.

Although regarded as chemically inert, the cadmium sulfide pigments can react with $Cu(+2)$ during melt processing to yield black CuS. This color contaminant results in yellows shifting green and reds turning brown. One potential source of $Cu(+2)$ ion is copper based heat stabilizer for polyamide.

Ultramarine Pigments

Chemically classified as "Sodium-Aluminum-Sulfo-Silicates", the ultramarine pigments represent a class of heat stable, light fast pigments. Unfortunately, their chemical resistance is limited - readily attacked by both acids and alkali - and they are therefore not recommended for extended outdoor exposure. A small selection of acid resistant grades protected with a silica coating are available at a slightly higher cost. A natural form of ultramarine is the mineral lapis lazuli.

The ultramarines range in color from a medium shade blue through violet to pink. A household example of an ultramarine blue in full shade is the "Milk of Magnesia" bottle. Standard grades of ultramarine pigments (uncoated) react with polyacetal resins, and I have observed color drift of these pigments in PC and PA 6/6. Some properties of ultramarine pigments are listed in Table 1.

ORGANIC PIGMENTS

Organic pigments range in complexity from simple carbon black to the heme-like structures of the phthalocyanine pigments (Figure 1). The use of organic pigments in polymer blends and alloys has been increasing rapidly as end-users and color formulators move away from heavy metal pigments. Typically, organic pigments have 10-20 times the color strength of inorganics of comparable hue. This is due to their much smaller particles size - in the

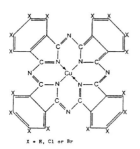

X = H, Cl or Br

Figure 1. The phthalocyanine ring system

range of 0.05 microns. On the one hand, the small particle size affords the possibility of creating transparent colors. On the other, it means that organic pigments are often very hard to disperse. On a per pound basis organic pigments are generally much more expensive than inorganics. However, because of their inherent strength they are sometimes more cost effective. In general, our experience has been that replacing heavy metal pigments with organic pigments leads to higher coloring costs.

Another down side in the use of organic pigments is potential reactivity with the resin. Many organic pigments contain the $C = 0$ (carbonyl) group as part of the chromophore (the structure within the molecule which confers absorption in the visible region). Polyamide resins are weak reducing agents which can convert carbonyl groups to hydroxyl groups. This reaction often results in the destruction of the chromophore and a complete loss of color. Care must therefore be taken in the selection of organic pigments (and dyes - see below) when coloring polyamide based blends and alloys.

Carbon Black Carbon black pigments are usually made by either of two processes. Most are produced by the furnace process in which petroleum is reduced to carbon at high temperature. Since there are small quantities of suspected carcinogen polynuclear aromatic hydrocarbons produced as by-products, "furnace black" does not comply with regulations for food contact applications. The natural gas process takes place at much lower temperature and the resulting pigment is found to comply with FDA regulations. The resulting pigment called "Channel Black" is also considerably more expensive.

Carbon black ranges in strength from "low jetness" utility grades (about 0.06 microns) to "high jetness" grades (as low as 0.015 microns), which includes some conductive grades. Several carbon black grades with particle size in the region of 0.02 microns are found to confer excellent weathering properties to just about any plastic material when present at 2 pph. Carbon black properties are summarized in Table 2.

Phthalocyanine Pigments In the blue-green region of the color spectrum the formulator's requirements are well met by the phthalocyanine pigments, which represent some of the best cost for performance values in the industry. As a class, these pigments, which were discovered serendipitously in an ICI research lab, share the following desirable properties:

- low cost
- high strength
- clean, bright hues
- good to excellent heat stability
- excellent light fastness
- good to excellent weathering
- transparency
- meet food contact regulations (with some exceptions)

Table 2. Organic pigments

Pigment	Strength	Heat stability	Weather resistance	FDA status	Cost $/lbs
Carbon black					
Furnace black (Pigment Black 7)	E	E	G-E	no	0.7
Channel black (Pigment Black 7)	E	E	G-E	yes	4.0
Phthalocyanines					
Red shade phthalo blue (Pigment blue 15:1)	E	G	G-E	yes	12.0
Green shade phthalo blue (Pigment Blue 15:3)	G-E	G	G-E	yes	10.0
Blue shade phthalo green (Pigment Green 7)	G	E	G-E	yes	15.0
Yellow shade phthalo green (Pigment Green 36)	G	E	G-E	no	18.0
Quinacridones					
Quinacridone red (Pigment Violet 19)	G	F	G-E	yes	30.0
Quinacridone violet (Pigment Violet 19)	G	F	G-E	yes	30.0
Quinacridone magenta (Pigment Red 122)	G	G	G-E	no	35.0
Quinacridone (Pigment Red 202)	G	G-E	G-E	no	40.0
Other organic pigments					
Disazo yellow (Pigment Yellow 93,95)	G	G	G-E	PY 95 yes	32.0
Quinophthalone yellow (Pigment Yellow 138)	G	G	F	no	27.0
Disazo red (Pigment Red 220)	G	G	G	yes	45.0
Azoic red (Pigment Red 177)	G	G	G	yes	60.0

These along with other properties of the phthalocyanine pigments are summarized in Table 2.

The structure shown in Figure 1 is the parent compound of the phthalocyanine pigments. The unsubstituted (no halogen atoms) ring system yields a blue pigment which is available in two crystal forms, alpha (red shade) and beta (green shade). The commercial pigments usually have some Cl substitution, which provide additional heat stability.

Chlorination of the ring system yields a blue shade green pigment which, in my experience, has the highest heat stability of any organic pigment (other than carbon black). Replacing some of the chlorine substituents with bromine atoms yields a yellow shade green.

The major problems associated with the use of phthalocyanine pigments in polymer blends and alloys are:

- phthalo blues are slightly soluble in polystyrene and may lead to color control problems in styrene based materials;
- all phthalo pigments may act as nucleators in crystalline resins - leading to uncontrolled shrinkage and warpage - and must be used with care in crystalline materials. (Note: the phthalo green pigments display the largest effects.)

Quinacridone Pigments First commercialized in 1958, the quinacridone, QA, pigments rapidly gained acceptance in the coatings industry as a replacement for heavy metal pigments due to their high strength, good light fastness and weathering properties, and moderate cost. Figure 2 shows the chemical structures of the three most commonly used QA pigments in plastics. These pigments, whose properties are summarized in Table 2, are in the red and violet range of the color spectrum.

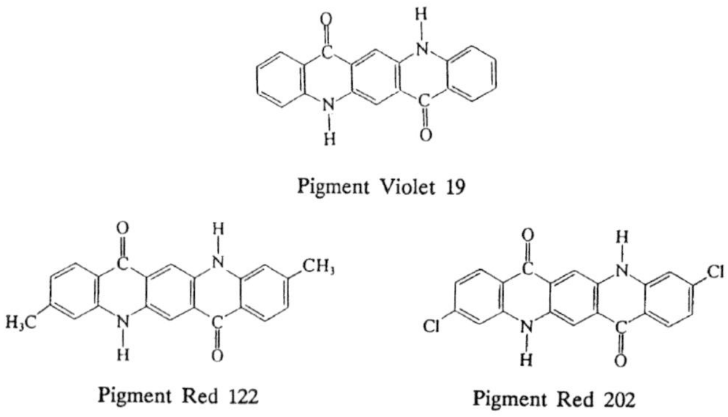

Pigment Violet 19 is polymorphic, and can be produced in a blue shade (beta crystal) that is a bright red shade violet and in a yellow shade (gamma form) that is a deep red. By adjusting the particle size QA suppliers can produce a variety of red and violet shades.

Pigment Violet 19

Pigment Red 122

Pigment Red 202

Figure 2. Quinacridone pigments.

The dichloro-QA (Pigment Red 202), a bright magenta pigment appears to have the broadest applicability in engineering resins and their blends and alloys. It has the highest heat stability (over 550F) and has the lowest tendency to solubilize during melt processing. Its one drawback is that it has not been cleared for food contact use. The other QA pigments, especially Pigment Violet 19, will solubilize to various degrees in PS-based materials and result in the appearance of fluorescent green streaks in the part. We have also seen this happen in PET. QA pigments react with polyamides, which are reducing agents, and lose their color. They are therefore not suitable in PA based blends and alloys.

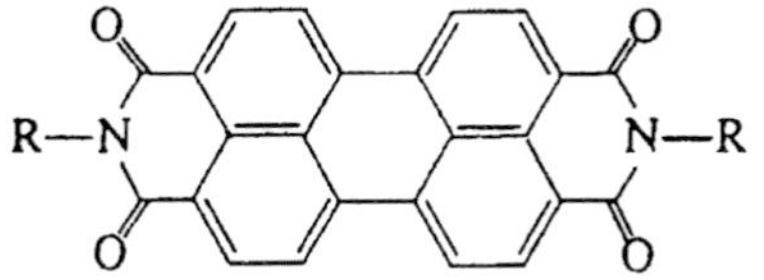

Figure 3. The perylene ring system.

Perylene Pigments Based on the ring structure shown in Figure 3, the perylene pigments represent a class of highly heat and light stable pigments in the red region of the spectrum. In general, these pigments are harder to disperse and more expensive than the QA reds and violets. For the past several decades, perylenes; have been used extensively in automotive coatings. Consequently, they are used more and more in automotive plastics applications.

Because of the increased heat stability, perylenes can be used in a wider variety of blends and alloys than the QA pigments. However, like the QA's, they are not recommended for materials containing polyamide resins.

Other High Performance Organic Pigments Table 2 lists a variety of other high performance organic pigments which have been used successfully in blends and alloys. Of these, the disazo condensation pigments are most noteworthy because of their broad clearance for food contact applications. I have not had success with any of these pigments in polyamide 6/6 or its blends and alloys.

DYES

Because dyes dissolve into the resin matrix during melt processing, dispersion is not an issue. As a rule, dyes have very limited solubility in polyolefin resins. Avoid their use in blends and alloys that have an olefin component. Some desirable properties which most high performance dyes share are:

- transparency
- excellent color strength
- ease of dispersion
- good light fastness in transparent colors

On the other hand, high performance dyes are expensive, have only fair color fastness (tend to migrate from the plastic), and generally have only poor to fair light fastness in opaque colors. In addition, dyes can have adverse effects on the plastic materials themselves. These include

1. acting as a plasticizer and reducing thermal properties
2. reactions. For example, some grades of the dye Solvent Yellow 163 contain residual chloride ion from an intermediate step and, therefore, cannot be used in polycarbonate (chloride ion promotes the degradation of PC).

The two major classes of dyes used in melt coloring polymeric materials are the anthraquinones and the perinones. Between the two they offer the colorist a complete range of colors from yellow to violet - and many are suitable for use in polyamide blends and alloys.

Table 3. Dyes

Dye	Strength	Heat stability	Weather resistance	FDA status	Cost $/lbs
Antraquinone dyes					
Solvent Red 111	G	E	E in mass F in tint	no	25.0
Solvent Red 52	G	E	G	no	100.0
Solvent Blue 59	G	F-G	F	no	35.0
Solvent Blue 97	G	E	F	no	55.0
Solvent Yellow 163	G	E	F	no	30.0
Solvent Green 28	G	E	F	no	65.0
Solvent Green 3	G	G	F	no	40.0
Solvent Violet 13	G	F	F	no	45.0
Perinone dyes					
Solvent Red 135	G	G	E in mass F in tint	yes, PET	35.0
Solvent Red 179	G-E	E	F	no	43.0
Solvent Orange 60	G	G	E in mass F in tint	no	40.0
Fluorescent dyes					
Solvent Yellow 160	E	G-E	F	no	100.0
Vat Red 41	E	G	F	no	85.0
Solvent Orange 63	E	G	F	no	115.0

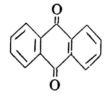

Figure 4. The anthraquinone ring system.

Anthraquinone Dyes

The anthraquinone dyes are based on the structure shown in Figure 4. Key properties are summarized in Table 3. Most commercial anthraquinone dyes have sufficient heat stability to be used in polycarbonate and thermoplastic polyesters. However, as indicated in Table 3, only a handful of these dyes are suitable for polyamide resins and their blends and alloys. Even in these cases caution must be applied. Polyamide materials colored with red anthraquinone dyes have been observed to slowly shift bluer over time. One hypothesis states the color shift is due to the slow absorption of moisture.

Perinone Dyes

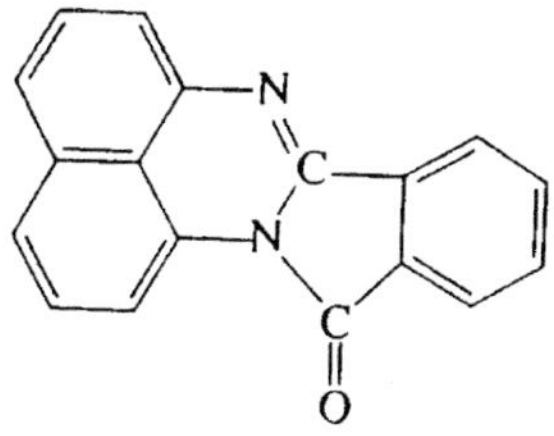

Figure 5. The perinone ring system.

Figure 5 shows the structure of Solvent Orange 60, a perinone dye. The properties of three important perinone dyes are listed in Table I SR 135 and SO 60 have been widely used in engineering resins for decades. Both are used, for example, in automotive lenses. SR 135 is used in both acrylic and PC "tail-light red" formulations, and SO Orange is the familiar "parking light amber". Because of their heat stability, both are finding use in a variety of engineered alloys and blends. SR 135 reacts with polyamides and is not recommended for NY6 or NY66 applications or in blends and alloys based on this resin. More recently, several producers have been offering SR 179, a perinone with superior heat stability and strength compared to SR 135. It holds up well in both PA6 and PA66, even under high shear injection molding applications.

Fluorescent Dyes

Fluorescent colorants convert energy from the near UV region of the spectrum to visible light. The result is that the colored object reflects more visible light than it receives and appears to glow in daylight. Fluorescent dyes are used widely in the packaging industry to attract the consumer to their products on store shelves. The majority of fluorescent dyes (on a weight basis) goes into the production of fluorescent "pigments", which are solutions of the dyes at 2-5% concentration in an oligomer carrier. These "pigments" are then compatible with polyolefins such as HDPE (for bottles) and PPRO (for caps). Some grades have sufficient stability for blends and alloys based on olefin and styrenic resins. A small sub-class of these dyes, however, has the heat stability required for engineering resins and their blends and alloys. Again, one cannot use heat stability alone as a guide to which materials can be colored by a given fluorescent dye. Reactivity rears its head. SO 63, e.g., is unsuitable for polyamides.

The structure of Solvent Green 5, a bright primrose yellow dye, is shown in Figure 6, and the properties of three popularly used high performance fluorescent dyes are given in Table 3. All of the fluorescent dyes suitable for high heat applications share high color and tinting strength, intense chroma, and outrageous pricing. The range is from $75-125/lb. Some special dyes with extraordinary brilliance are being offered at about $3000/lb.

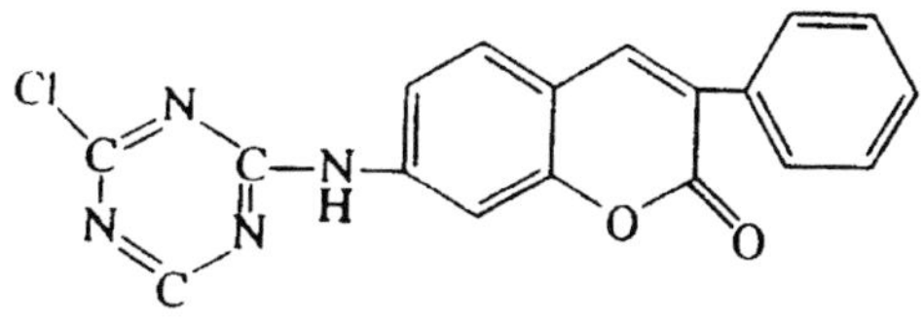

Figure 6. Solvent Green 5 (a fluorescent dye).

What would motivate a colorist to use these expensive dyes in a blend or alloy? Very often the material being colored is very opaque or has an inherent "dirtiness". Often, fluorescent dyes offer the only means to overcome these inherent characteristics to yield a clean, bright color.

CONCLUDING REMARKS

Make note of this: COLOR SELLS! If you want a new high performance thermoplastic alloy or blend to reach the widest number of appropriate end user markets, you have to be able to color it in a cost effective manner that does not harm its performance properties. Barriers to cost effective coloring include:
- the material's inherent color and opacity;
- chemical incompatibility with one or more polymeric or compatibilizer components;
- physical incompatibility with one or more polymeric components (many materials will not physically accept dyes, e.g.);
- stringent heat stability and/or weathering requirements.

Of these barriers, the one that is most overlooked is the first. Many of the new thermoplastic materials coming into the market place are blends and alloys that are specifically engineered to provide a combination of the properties of the individual polymers. Often these materials combine crystalline and amorphous polymers with an impact modifier. The products of these marriages often contain a maze of phase boundaries that result in light scattering (milkiness) equivalent to as much as 0.5% titanium dioxide. Obtaining high chroma colors (e.g., some electrical code colors or even a jet black) in the presence of this inherent milkiness becomes an expensive proposition. Often so much color has to be added to the material formulation that critical material properties are affected - a double whammy, cost and performance.

Photochromic Dyes of Enhanced Performance

David Clarke, Fiona Ellwood, James Robinson
Keystone Aniline Corporation, Chicago, IL, USA

Photochromic compounds are those which can be reversibly transformed between two states having different absorption spectra, such change being induced in at least one direction by the action of electromagnetic radiation (e.g. ultra-violet light).

$$A(\lambda_1) \xleftrightarrow{\ hv\ } B(\lambda_1);\ A \xrightarrow{\ hv_1\ } P;\ P \xrightarrow{\ \Delta/hv_2\ } A\ \textit{(unimolecular reactions)}$$

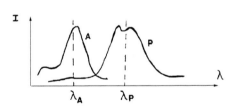

Figure 1. Reversible transformation.

Photochromic compounds are not new, being first reported in 1807 by Fritsche. Since that time a large number of inorganic and organic systems showing photochromic properties have been reported.

Existing technology in this area includes some products already in the market place such as the silver halide glass based Reactolite Rapides and some simpler organic photochromic dyes.

The potential use for photochromism, in marketable products, is huge with applications in plastic lenses, imaging, agriculture, fashion, optical storage, advertising, military, security, and novelties.

Organic photochromics exhibit good coloration and, unlike the silver halides, can be incorporated into plastics, widening the fields of possible use.

This paper will deal briefly with the properties and synthesis of spiroindolinonaphthoxazines (SINO's) and chromenes, and review some of the optical and performance properties.

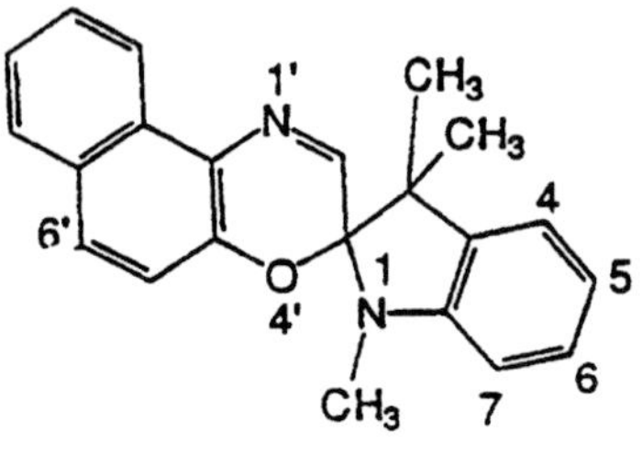

Figure 2. Chemical structure.

Spiroindolinonaphthoxazies were first patented in 1959, but the early products had a limited color range (550-620 nm), had poor colorability (induced optical density - IOD) suffered from fatigue (loss of colorability) and they tended to have a base color.

The general method of a SINO synthesis is to condense a 1-nitroso-2-naphthol with a 2-methyleneindoline. Variants of the SINO system can be produced therefore by altering the substituents in the nitrosonaphthol, or by varying the nature of the methyleneindoline.

In investigating the effect of varying substituents, it became apparent that improvements to the performance of the SINO were possible. The introduction of an amino substituent into the 6' position of the nitrosonaphthol resulted in a blue shift in the derived SINO and this was also accompanied by an increase in intensity of color in the darkened state.

6' substituent	λ_{max}, nm
H	605
Indolio	605
Piperidino	578
Morpholino	580
Aziridino	574
Diethylamino	574

The blue drift can also be enhanced by the introduction of electron withdrawing substituents, at C-5.

6'	C-5	λ_{max}, nm
Piperidino	H	578
Piperidino	Cl	568
Piperidino	CF_3	560

Figure 3 shows the ring closed (bleached) structure and the ring opened (darkened) structures.

The introduction of an amino substituent into the 6' position results in greater absorption in the UV region, leading to increased photon capture and improved quantum efficiency for darkening (Figure 4).

Figure 5 shows the increase in IOD induced optical density, due to the presence of an amino group in the 6' position.

Because of the increased absorption in the 350-420nm region, selected UV absorbers can be incorporated into the system which will confer stability whilst allowing the photochromic effect to take place.

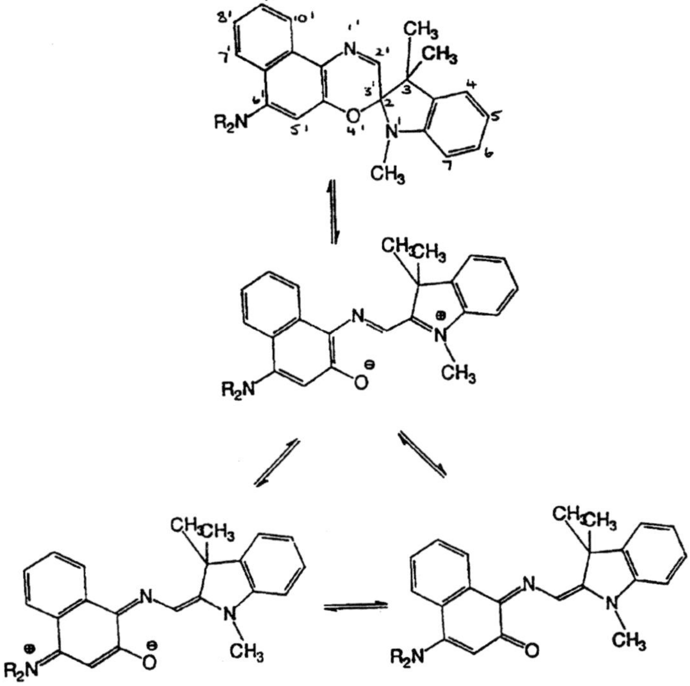

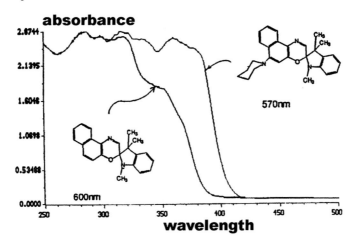

Figure 3. 6'-amino SINO photochromics.

Figure 4. Change of UV spectrum on substitution.

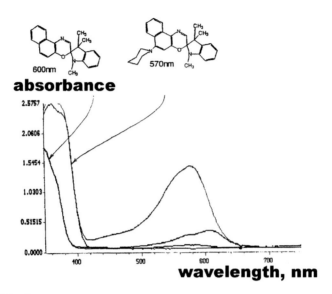

absorbance

wavelength, nm

Figure 5. Induced optical density.

The effect of introducing an amino substituent into the 6' position is:

2	X	Coefficient of absorbance
3	X	Photon capture
5	X	Quantum efficiency

resulting in dense coloring dyes, allowing lower product loadings.

Investigation into the efforts of introducing a nitrogen hetero atom into the indoline ring to give a 7 aza and a 4,7 diazo SINO revealed a large hypsochromic shift accompanied by an increase in stability. The products were colorless in the bleached state.

Aza Substitution Effects

Water white bleached states: Colorless fabricated articles

Hypsochromic shift

Hypsochromic effect: Cosmetically coloring materials

Increase in stability

Varying the N^1 alkyl substituent in a bid to improve solubility, resulted in an unexpected and beneficial effect. The fade rate, the rate at which the darkened state returns to the bleached slats, was found to vary with the change in alkyl group. This effect was also noticed at higher temperatures.

Varying the N-alkyl substituent had no effect on the color only on the IOD at higher temperatures and the fade rate. Dyes can thus be used in mixtures to tailor the fade rate so that

7 AZA

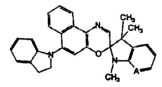

4,7 DIAZA

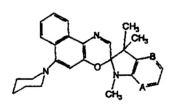

A	$\Delta\lambda_{max}$	λ_{max} (PU)
CH	0	606
N	-34	572

A	B	$\Delta\lambda_{max}$	λ_{max} (PU)
CH	CH	0	578
N	CH	-38	540

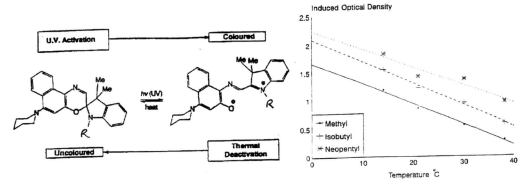

Figure 7. Induced optical density vs. temperature.

R	ODF10	IOD			
		14C	21C	30C	38C
Methyl	29	1.19	0.85	0.55	0.29
Isobutyl	17	1.56	1.22	0.94	0.58
Ethylhexyl	15	1.49	1.28	0.97	0.63
Phenylpropyl	13	1.43	1.25	0.95	0.55
Neopentyl	9	1.84	1.43	1.37	0.96

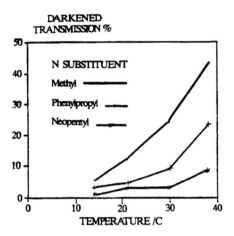

Figure 8. Temperature sensitivity of N-substituted 6' amino SINO derivatives.

color changes are minimized as the darkened state builds up and as the dyes bleach enabling the product to give the same performance across a temperature range.

There are no changes in fatigue performance.

ODFIO can be expressed as the percentage loss in color after 10 seconds fade time. There are observed changes in λ_{max} when the SINO's are incorporated into polymers or dissolved in solvent. The following table shows the effect of solvent on a 6' indolino SINO.

solvent	λ_{max}
cyclohexane	556
ethylacetate	583
dichloromethane	591
acetone	596

Measurement of the darkened state spectra is not easy and some assumptions have to be made due to the temperature dependent backwards bleaching action that takes place. Bleached state properties are absolute, but the darkened state properties are dependent on:

1   the power of the illuminating source
2   spectral output of the illuminating source
3   temperature

Comparisons should be made at the same time and standards included in the test.

The extent of fatigue that a photochromic dye undergoes is directly related to the energy in the form of UV light that it has received. It is independent of the number of times the dye ring opens and ring closes.

There are several fatigue reactions which can occur: water, oxygen, including singlet oxygen and free radicals, being the main problems. Photochromic dyes do however have an intrinsic fatigue rate, even when all oxygen, water and solvents are removed. Fatigue demonstrates itself as a reduction in colorability, loss in IOD, and an increase in yellowness of the bleached state.

Hindered amines are useful stabilizers, but they themselves have a tendency to yellow, and UV absorbers.

Work on the performance of photochromic dyes has led to a range of products with improved performance and dye strength with λ_{max} values in the range 430-632 nm.

The performance of the dyes depends on the delivery systems, the pH, nature of the host polymer, and the method of application. Direct casting of photochromics into acrylics, polyurethanes and polyolefins has been achieved. Barred temperatures of up to 260°C have been employed with little degradation of the photochromic dye taking place.

Further work on the stability and methods of incorporation of photochromic dyes into plastics is required in order to widen their use.

US Patents 4913544 and 4851530. Application 08/160184, and GB Applications 93/168904, 93/168573 and 93/168581, cover the products referred to in the paper.

Three Color Effects From Interference Pigments

Louis Armanini, Henry L. Mattin

The Mearl Corporation, Ossinging, N Y, 10562, USA

INTRODUCTION

Everyone by now is aware of pearlescent and interference pigments. Although not a household word, they are used extensively in every imaginable application. They are incorporated in plastics, coated on substrates, used in printing inks, in cosmetics and used extensively in automotive paints.

Although pearlescent pigments encompass the natural pearl essence derived from the fish and also the various crystal forms of bismuth oxychloride, it is the coatings of titanium dioxide and iron oxide on mica that have gained the most acceptance and are the most commercially successful. It is the coatings of titanium dioxide on mica and their derivatives that are the subjects of this study.

When mica platelets are coated with a smooth thin uniform layer of titanium dioxide and processed, a lustrous, highly reflecting pigment results which reflects light across the visible spectrum. If the thickness of the titanium dioxide layer is increased, interference of visible light takes place, and interference colors and pigments are produced. When absorption colorants are precipitated upon interference pigments, interesting color effects are produced. Three colors can be observed depending on the colorants, the interference pigments, the background and the viewing angle. On a transparent substrate one color is seen at the specular angle, another color at the diffuse angle, and a third color may be seen by transmission. Thus three distinct colors may be seen from one pigment. It is these color effects which are the subject of this paper.

In order to understand how three distinct colors from one pigment are possible, it is first necessary to understand the optical properties of pearlescent and interference pigments and also the optical properties of absorption colorants. The manner in which these reflect light differs and this forms the basis of the three color effect.

OPTICAL PROPERTIES

ξments reflect light in a specular manner similar to a mirror. ...ight is reflected is equal to the angle of the incident light. The reflection is highly directional with most of the light being concentrated at the specular angle. Unlike a mirror, however, where all of the light is reflected, pearlescent pigments reflect only part of the light. That part which is not reflected is transmitted to the next layer where further reflection occurs. The amount of light which is reflected specularly depends on the index of refraction and the smoothness of the surface. The higher the index of refraction and the smoother the surface, the greater will be the reflection from each surface. A diagram showing the reflection of a pearlescent pigment is shown in Figure 1.

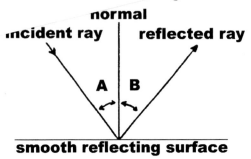

Figure 1. Specular reflection of pearlescent pigments.

Absorption colorants do not reflect in a specular manner. They reflect light mostly by scattering. Their surfaces are not smooth but irregular, and absorption, reflection and scattering of light take place. Comparison of the reflection of a pearlescent pigment with that of an absorption colorant is seen in Figure 2.

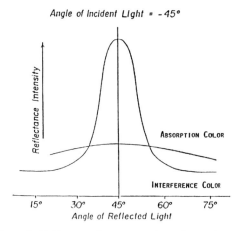

Figure 2. Reflectance from a smooth surface and a matte surface.

If light is allowed to fall upon a smooth surface similar to a pearlescent platelet at an angle of 45 degrees from the normal, and if a detector is placed at an angle of 45 degrees away from the normal, and the intensity of the reflected ray is recorded as a function of the reflected angle, a pearlescent pigment produces a response similar to that shown. Most of the reflected light is concentrated at an angle of 45 degrees and a few degrees on either side of it. At angles away from the specular, very little light is reflected.

If an absorption colorant or a matte or irregular surface is so examined, the light is not reflected at the specular angle but is reflected at all angles. This is the fundamental difference in reflectivity between a pearlescent pigment and an absorption pigment: The amount of light which is reflected and the manner in which it is reflected.

TITANIUM DIOXIDE COATINGS ON MICA

PEARLESCENT PIGMENTS

Mica platelets can be coated with a thin, very smooth uniform layer of titanium dioxide. If the thickness of the titanium dioxide is kept at about 60 nanometers, the platelets reflect light across the visible spectrum from 400 nanometers to 700 nanometers. A diagram showing a mica platelet coated with titanium dioxide is shown in Figure 3.

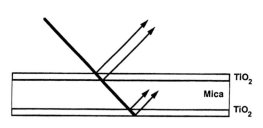

Figure 3. Titanium dioxide coating on mica.

There is actually reflection at the four interfaces or where there is a change in index of refraction from a high index to a low index or from a low index to a high index.

INTERFERENCE PIGMENTS

If the thickness of the titanium dioxide layer is increased beyond the 60 nanometers of the white reflecting pigments, interference of light takes place. The ray which is reflected from the first surface of the TiO_2 interferes with the ray from the second surface. It can interfere constructively or destructively. The result is that interference pigments are produced whose colors depend solely on the interference of light. White light is therefore separated into two colors or components. One color is seen by reflection at the specular angle, and its complementary color is transmitted through the platelet and is seen by transmission.

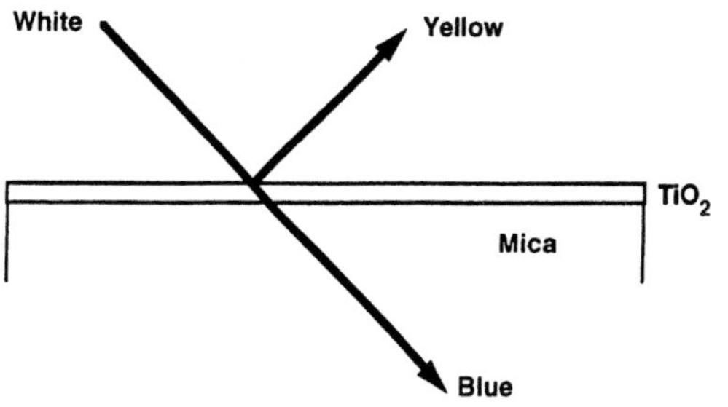

A diagram showing the reflection of a yellow reflecting interference platelet is seen in Figure 4.

At the specular angle by reflection we see a yellow, and by transmission its complementary color which is blue. Since the color depends on the thickness of the titanium dioxide layer, a large number of colors are possible. Also second and third colors are also possible.

Figure 4. Yellow reflecting interference platelet.

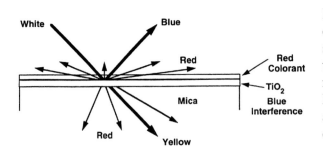

Figure 5. Red absorption pigment coated on a blue reflecting interference platelet.

COMBINATION PIGMENTS

In order to enhance the reflection color of interference pigments, absorption colorants are precipitated on the surfaces of the interference pigments. These are called Combination Pigments. Different colors are seen at different angles. The reflection color of the TiO_2 interference layer is seen at the specular angle since its reflection is specular. At diffuse angles where the specular reflection is no longer dominant, the color of the absorption pigment can be observed. Its reflection is not specular but diffuse. In Figure 5, a red colorant which has been coated on a blue reflecting TiO_2 coated mica pigment is shown.

At the specular angle we see the blue from the interference pigment. At the diffuse angle we see the red from the absorption colorant which has been precipitated on the surface. By transmission, a third color is formed which is not often seen unless coated on a transparent substrate such as glass or Mylar. The transmission color from the interference pigment which is yellow has combined with the absorption colorant which is red to form a third color which is orange.

The combination of absorption colorant and interference pigment has resulted in a third color being formed. We will, therefore, examine the various combinations of colorants and interference pigments which are possible and the results which they produce.

It must be understood that the concentration of colorant and the concentration of the interference pigment must be adjusted so that the reflection intensities are approximately the same. If not, one color will dominate over the other, and color changes will not be seen.

INTERFERENCE PIGMENTS AND ABSORPTION COLORANTS USED

The four basic interference pigments which were used and upon which the absorption colorants were applied reflect yellow, red, blue or green. These pigments are identified on the chromaticity coordinate diagram in Figure 6. As can be seen the pairs of colors are virtually complementary. A straight line connecting the yellow and the blue pigments passes through the neutral C point. The proper mixture of the yellow and the blue pigments will yield a neutral or white pigment since colors produced by interference of light combine by an additive process. If they combined by a subtractive process as occurs when absorption colorants are mixed, green would be formed. The red and green pigments are almost complementary. A

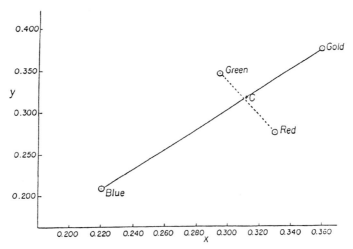

Figure 6. CIE chromaticity coordinates of interference pigments.

straight line passing through the two points does not pass through the C point, but is slightly adjacent to it.

The transparent colorants which were used were a yellow (Fe_2O_3), red (Carmine), blue (iron blue) and green (Cr_2O_3). Other colorants such as phthalocyanine blue, phthalocyanine green, quinacridone red can be substituted for these pigments and actually such pigments are available.

Generally the concentration of all the colorants was approximately 2% to 5% by weight based on the mica platelets. At these concentrations the color intensity of the colorants was approximately equivalent to the reflection intensity of the interference pigments.

In order to study the various colors produced, the combination pigments were incorporated into a nitrocellulose lacquer and coated on glass slides. The colors were observed visually and some were analyzed using a goniospectrophotometer. The glass slides were prepared in the following manner.

PREPARATION OF SLIDES

The dry pigments were dispersed at a concentration of 3.0% in a nitrocellulose lacquer having a solids content of 9.5%. Films of the dispersed particles in the nitrocellulose lacquer were formed on glass slides using a Bird Film Applicator. This forms a wet film of approximately 0.003 inch. The films were allowed to dry so that the platelets were aligned parallel to the film. The films coated on the glass slides were then analyzed using a Leres Trilac Goniospectrophotometer which measures reflection and transmission from 400 to 700 nanometers. This instrument is described in Reference 2.

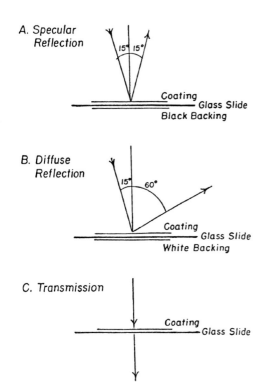

A. Specular Reflection

15° 15°

Coating
Glass Slide
Black Backing

B. Diffuse Reflection

15° 60°

Coating
Glass Slide
White Backing

C. Transmission

Coating
Glass Slide

Figure 7. Reflection of light by interference pigments coated on glass slides.

Monochromatic light falls on a sample specimen at a desired angle of incidence. In this work, the angle of incidence which is used is 15° from the normal written by convention as -15°. The light is reflected and analyzed by a suitable detector. The angles of reflection or "viewing" which are used are 15° and 60° from the normal. These conditions are written as -15/15 for specular reflection and -15/60 for diffuse reflection.

A reference beam is used with a plate of $BaSO_4$ as the reference standard. Thus, all measurements are relative and are compared to the reflection from a barium sulfate standard. Because the specular reflection from a pearlescent pigment is so much greater than the reflection from a barium sulfate plate, a neutral density filter allowing only 10% of the light to pass was placed in the beam incident on the sample for measurements made at -15/15.

When measurements were made at the diffuse angle of -15/60, the amount of light reflected was quite small, and the neutral density filter was removed. Therefore, all measurements at -15/15 were 10 times the intensity of those made at -15/60.

When measurements were made at -15/15, a black card was placed behind the coated glass slides to eliminate the effect of the transmission color. When measurements were made at -15/60, a white card was placed behind the coated glass slide. A diagram demonstrating the position of the slides and the cards is shown in Figure 7.

When transmission measurements were made, the glass slides were placed in the beam path at perpendicular incidence, and the signal increased electronically by a factor of 2. The result was that the transmission measurements were one-fifth those made at -15/15.

There are a number of combinations of absorption colorants and interference pigments which are possible. The combinations which will be examined are the following:

I Absorption Colorants and Transmission Colors are Different
II Absorption Colorants and Transmission Colors are the Same
III Absorption Colorants and Transmission Colors are Complementary

COLORANTS AND TRANSMISSION COLORS ARE DIFFERENT

Table 1. Colorant and transmission colors are different

Colorant added	Reflection color	Transmission color	Third color
Yellow	Red	Green	Yellow-Green
Yellow	Green	Red	Orange
Red	Yellow	Blue	Violet
Red	Blue	Yellow	Orange
Blue	Red	Green	Blue-Green
Blue	Green	Red	Violet
Green	Yellow	Blue	Blue-Green
Green	Blue	Yellow	Yellow-Green

In Table 1, examples are shown where the colorant which was added to the interference pigment had a color which was different from the transmission color of the interference pigment. In all these cases a new third color was formed.

In the first group a yellow colorant was added to red and green reflecting interference pigments. Yellow-green and orange third colors were observed respectively.

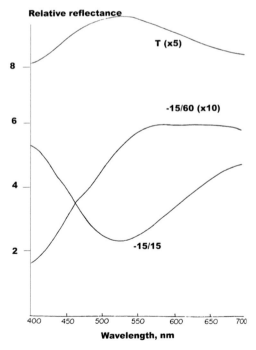

Figure 8. Spectrophotometric curves of a yellow absorption pigment coated on a red interference pigment at -15/15, -15/60 and by transmission.

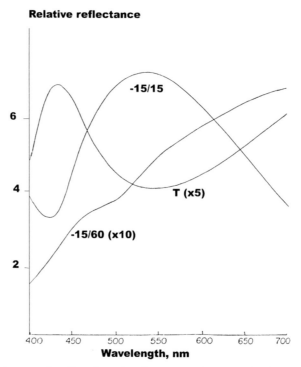

Figure 9. Spectrophotometric curves of a yellow absorption pigment coated on a green interference pigment at -15/15, -15/60 and by transmission.

In the next group a red colorant was added to both yellow and blue reflecting interference pigments resulting in violet and orange third colors being formed.

When a blue colorant was added to red and green reflecting interference pigments, blue-green and violet third colors were formed respectively.

In the last group a green colorant was added to both yellow and blue reflecting interference pigments, and blue-green and yellow-green third colors resulted.

The third colors which are shown in the Table 1 were observed visually. Goniospectrophotometer curves were obtained and they are shown in Figures 8, 9, 10 and 11 and 12.

COLORANT AND TRANSMISSION COLORS ARE THE SAME

In Table 2, examples are shown where the colorants added to interference pigments had the same color as the transmission color. An enhancement of that color resulted as may be expected. A reinforced yellow, red, blue and green resulted.

TABLE 2. Colorant and Transmission Colors are the Same

Colorant added	Reflection color	Transmission color	Third color
Yellow	Blue	Yellow	Reinforced yellow
Red	Green	Red	Reinforced Red
Blue	Yellow	Blue	Reinforced blue
Green	Red	Green	Reinforced green

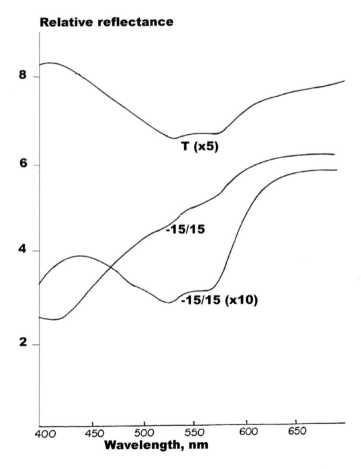

Figure 10. Spectrophotometric curves of a red absorption pigment coated on a yellow interference pigment at -15/15, -15/60 and by transmission.

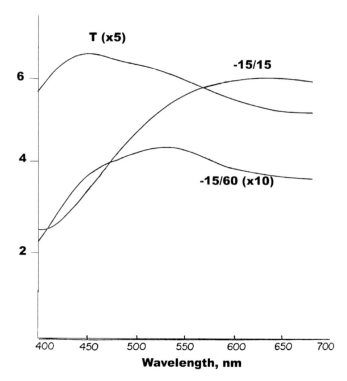

Figure 11. Spectrophotometric curves of a green absorption pigment coated on a yellow interference pigment at -15/15, -15/60 and by transmission.

COLORANT AND TRANSMISSION COLORS ARE COMPLEMENTARY

In Table 3, examples are shown where the color of the colorant which was added was complementary to the transmission color of the interference pigment. Some interesting results were obtained. The colors from pure interference pigments mix by an additive process. Thus a yellow and a blue when mixed in a 1:1 ratio yield a neutral white. Absorption pigments on the other hand mix by a subtractive process. Mixing a yellow pigment and a blue pigment generally results in a green pigment.

As can be seen in the table, the first two examples yielded a neutral white. The yellow colorant combined with the blue transmission color resulting in a neutral white. The red colorant combined with the green transmission color resulting in a neutral white.

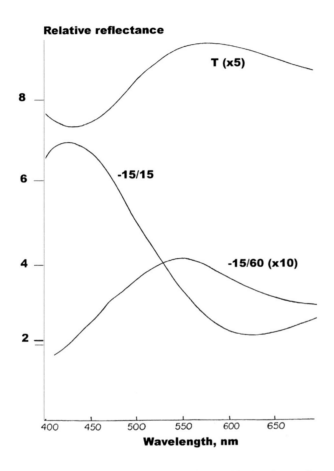

Figure 12. Spectrophotometric curves of a green absorption pigment coated on a blue interference pigment at -15/15, -15/60 and by transmission.

Table 3. Colorant and transmission colors are complementary

Color added	Reflectance color	Transmission color	Third color
Yellow	Yellow	Blue	Neutral
Red	Red	Green	Neutral
Blue	Blue	Yellow	Blue-Green
Green	Green	Red	Orange

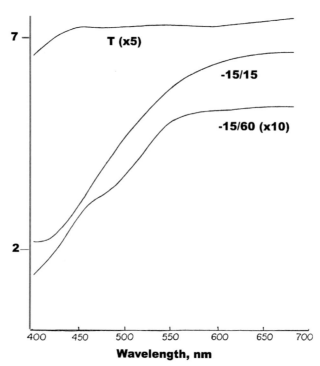

Figure 13. Spectrophotometric curves of a yellow absorption pigment coated on a yellow interference pigment at -15/15, -15/60 and by transmission.

In the last two examples in the table, the blue colorant combined with the yellow transmission color resulting in a blue-green color and the green colorant combined with a red transmission color to yield an orange.

A goniospectrophotometric curve of a yellow absorption pigment coated on a yellow interference pigment is shown in Figure 13 . At the specular angle of 15/15, the interference yellow is observed. At 15/60 the yellow colorant is seen and by transmission a neutral white is observed.

The colors observed by transmission are the principal colors. Slight changes in color away from the principal color are observed if the slides are tilted from the normal. This is due to the fact that interference colors shift to lower wavelengths with increased angle of viewing.

SUMMARY

The optical properties of pearlescent and interference pigments were examined and compared to absorption colorants. The reflection from pearlescent and interference pigments is specular. The reflection from absorption pigments is scattered and diffuse. Because of this difference, interesting color effects are observed when the two are mixed. When absorption colorants are directly precipitated on interference pigments, combination pigments are formed. Three distinct colors can be observed depending on the colorants and the angle of observation. At the specular angle the reflection color from the interference pigment is observed. At the diffuse angle or angles away from the specular, the absorption colorant is seen. A third color can be seen by transmission which can be entirely different from the reflection color or the absorption color. The absorption colorant mixes with the transmission color of the interference pigment to form a new third color, an enhancement of the absorption color or a neutralization of color depending on the colors and how they mix.

REFERENCES

1. L. M. Greenstein, **Pearlescence. Pigment Handbook**, Vol. III, T.C. Patton, Ed., *John Wiley & Sons, Inc.*, N.Y. 1974, 357-390.
2. L. M. Greenstein, R. A. Bolomey, *J. Soc. Cosmet. Chem.*, **22**, 161-177 (March 4, 1971).
3. F. W. Billmeyer, M. Saltzman, **Principle of Color Technology**, 2nd Edition, *John Wiley & Sons, Inc.* 1981.
4. L. Armanini, *Paint Coat. Ind.*, **5**(8), (November, 1989).

Fluorescent Pigments as Plastic Colorants: An Overview

Darren D. Bianchi
Radiant Corporation, 2800 Radiant Avenue Richmond, CA, USA

INTRODUCTION

Fluorescence is a process of photo-luminescence by which light of short wavelengths, either in the ultraviolet or the visible regions of the electromagnetic spectrum, is absorbed and re-radiated at longer wavelengths. The re-emission occurs within the visible region of the spectrum and consequently is manifested as color.

The commercial development and sale of fluorescent pigments and colorants dates back to the 1940's in the field of graphic arts. Development was initially centered around the application of point-of-purchase displays, advertising, safety and identification. To date, fluorescent materials have gained widespread acceptance in a myriad of applications, including toys, fashions, and packaging.

Fluorescent pigments are often used in specific applications where a particular appeal is desired. Studies have been conducted with children and adults showing that fluorescent products are noticed earlier and seen longer than their conventional counterparts. As a result, designers have incorporated the use of fluorescent products in many creative ways to enhance product sales.

The unique brightness of a fluorescent may be employed alone when one is trying to set their product apart from the rest in a competitive situation. In addition, fluorescents can be used as an accent in contrast to a more drab color, or they may be added to conventional pigments to brighten an otherwise dull color.

Due to the specialty of this market, only three domestic and four foreign manufacturers have enjoyed any real success in the manufacture of fluorescent colorants.

NATURE OF FLUORESCENT PIGMENTS

As those who have processed fluoresce know, they differ significantly from conventional pigments not only in color but in chemistry as well. Conventional pigments can be organic or inorganic substances and are of extremely limited solubility. Their dispersion is usually achieved and enhanced by the application of shear and the pigment particles tend to be more opaque in nature.

Daylight fluorescent pigments, however, are comprised of a solid state solution of fluorescent dyes in a friable polymeric resin. Once the dyes are incorporated into the resin, they are ground into a fine powder for use as a pigment or colorant. As these pigments are resinous solutions of dyes, they tend to be transparent in nature.

MANUFACTURING PROCESSES

The techniques employed to make the fluorescent pigments have varied over the years. The original method used was the bulk polycondensation reaction of melamine, formaldehyde, and toluene sulfonamide. The resulting products were tailored to various applications by being thermoplastic or thermoset depending upon the mole ratios of the polymer's raw materials.

As the fluorescent products evolved and certain shortcomings of the above products were noted, similar bulk condensation polymerization methods were carried out to make polyesters[1] and polyamides.[2] These are currently the most widely used fluorescent colorants for plastics. The polyesters allow for lower processing temperatures (<400°F) for bright, clean colors while the polyamides allow for higher processing temperatures (>400°F) and greater shear to achieve color development.

Another method developed in the 1970's to manufacture a pigment similar to the first by using suspension polymerization.[3] This technique offered pigments which combined bright colors and excellent inertness. This was due to the high degree of polymerization which was achievable in the droplet state. Until recently, this technique yielded colorants with poor color strength and high price and were thus not welcomed into the marketplace.

Another important component of fluorescent colorants are the dyestuffs used. Fluorescent pigments, as noted, are solid solutions of fluorescent dyes which do not fluoresce in an undissolved state. The dyes used in are predominately rhodamine (magenta) and coumarin (yellow) types. The fluorescent spectrum from yellow to magenta is achieved by combining the dyes at different ratios. The popular green color is made by adding yellow dye and phthalo green to the resin carrier. The blue is a combination of phthalo blue and optical brighteners of the benzopyranone type.

ENVIRONMENTAL CONSIDERATIONS

We have seen the recent rise of environmental legislation regarding packaging materials. As a result, designers and users are required to be increasingly more selective about which materials they use as they incorporate their awareness of environmental issues.

Most, if not all, fluorescent pigments do not contain any of the Coalition of North Eastern Governors (CONEG) heavy metals (cadmium, lead, mercury, and hexa-valent chromium). In fact, to the author's knowledge, no heavy metals are used in the manufacture of any fluorescent colorants.

In addition, many of the fluorescent products have been tested for skin irritation and acute toxicity. As a result, those that have been tested are classified as "essentially non-irritating" with a Draize Score of 0, and "essentially non-toxic" with an oral LD_{50} (rat) >5000 mg/kg.

QUALITY CONTROL

The quality control testing of fluorescent colorants by the manufacturers has been based upon attempted simulation of the compounders testing. The colorants are dispersed by either an injection molder, extruder, or a Banbury mixer into HDPE for observation in both mass tone and tint forms. In addition, a common method of display is a side-by-side pressout as performed on a hydraulic press. Carefully trained technicians perform visual observations while those in the fluorescent industry await the development of technology which will allow for adequate computerized color measurement.

INCORPORATION INTO PLASTICS

Once the fluorescent colorant passes quality control testing, it is then distributed to compounders to be made into color concentrate. Once in solid masterbatch or liquid concentrate form, the fluorescents are used in a wide variety of applications, including injection molding, rotational molding, blow molding, extrusion, and vacuum forming. These fluorescent colorants are used primarily in polyolefins, in vinyl plastisols, and somewhat less in styrenics, acrylics, and ABS.

By use of these methods and materials, one is allowed to create products for the toy, detergent bottle, traffic cone and safety equipment markets, among others. There is no a great deal of work done with fluorescents in film or fiber applications due to the inherent transparency and the relatively low tinting strength of these materials.

PROCESSING CHALLENGES

Some of the challenges that processors may face when handling fluorescent colorants are those such as the occurrence of plate-out. This phenomenon occurs when organic material, such as oligomeric species or fluorescent dyestuffs, thermally decompose and separate from the compounding mixture. Thus, these materials deposit on screws and other metal processing equipment.

Steps have been taken to address this problem by both the colorant manufacturers and the compounders. Manufacturers have worked to shift the molecular weight distribution of the colorants and reduce the lower molecular weight species.[4] Thus, reducing the likelihood of separation and decomposition. Compounders and additive suppliers have developed additive packages to reduce plate-out.

Other challenges in processing fluorescents may be heat instability or incompatibility of the colorants with the various resins. The manufacturers of fluorescent colorants are continually looking at ways to improve their processability in these regards.

CONCLUSIONS

The use of fluorescent pigments and colorants in plastics applications has experienced sustained growth over the past few decades. Research and development efforts continue in the pursuit of more thermally stable, plate-out resistant fluorescent colorants with greater tinting strength and opacity. As these qualities are achieved and improved upon, we should see the use and growth of fluorescent pigments and colorants into more diverse plastics applications well into the future.

REFERENCES

1   **U.S. Patent No. 3,922,232**, Alan K Schein, November 25, 1975.
2   **U.S. Patent No. 3,915,884**, Zenon Kazenas, October 28, 1975.
3   **U.S. Patent No. 3,945,980**, Tsuneo Tsubakimoto, Iwao Fuzikawa, March 23, 1976.
4   **U.S. Patent No. 5,094,777**, Thomas C. DiPietro, March 10, 1992.
5   **International Patent No. WO 9310191**, Kenneth Wayne Hyche, May 27, 1993.

Color Styling with Genuine Metallics in Plastics

Henning Bunge

Obron Atlantic Corporation, 27 Corwin Drive, PO Box 747, Painesville, OH 44077, USA

Aluminum and bronze pigments offer a wide range of unique coloristic effects, which in relation to their different optical characteristics will be described. In the past, the chemical resistance of metallic pigments have always been a problem. During the last few years various modified versions of both aluminum as well as bronze pigments have been developed which open new areas for their application in plastics. Their characteristics and advantages will be discussed and how to best incorporate metallic pigments into the different plastic resins.

If you ever had the opportunity to use aluminum pigments in plastics, you are familiar with the fact that you are not dealing with spherical pigments, but with flakes. Because of their shape, metallic pigments appear to behave in regard to their optical properties in almost every way opposite to non-metallic pigments. Here are some examples:

When reducing the particle size of organic colorants, the shade becomes brighter and the transparency increases. In contrast, metallic pigments with decreasing particle size hide better but display reduced brightness. To reduce the particle size of organic pigments, they have to be ground, sometimes at a high energy input. Metallic pigments must be incorporated gently to make sure the flakes are not damaged, which would cause the brilliance to be reduced.

How can this discrepancy between spherical non-metallic and the metallic flakes be explained? Metallic pigments reflect most of the light, scatter some but absorb very little, while particularly colored pigments absorb a large portion of the visible spectrum.

As illustrated the light reflects from the flat aluminum surface and scatters from its edges. The more light that is reflected, the brighter the appearance of the metallic shade, but with increased scattering, the brightness is reduced proportionally.

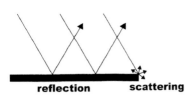

Figure 1. Reflection and scattering by aluminum flake.

Therefore, the brightest metallic effect is achieved with the largest flakes, because at an equal surface area, they have fewer edges than small flakes. The size of the largest flake is about 60 μm and the smallest is 2 μm.

Another way to reduce scattering is by using round flakes, illustrated by the SEM picture. The round edges scatter the light less because of a reduced ratio of surface to circumference and display the brightest metallic effect of all aluminum pigments.

Figure 2. Comparison of aluminum flakes. Left - round, right - normal.

The additional factor in achieving optimum brightness is to keep the particle size distribution as narrow as possible in order to reduce the amount of small particles present.

Table 1. Aluminum Pigments Overview

Flake type	Particle size distribution	Application	Optical appearance
Standard	Broad	Efficient hiding	Good hiding, darker
Standard	Improved	Polychromatic sparkle	Increasing brightness with increasing uniformity
Round	Improved	High brightness polychromatic sparkle	Brightest

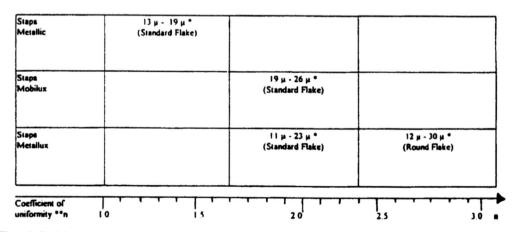

Figure 3. Particle size distribution characteristics.

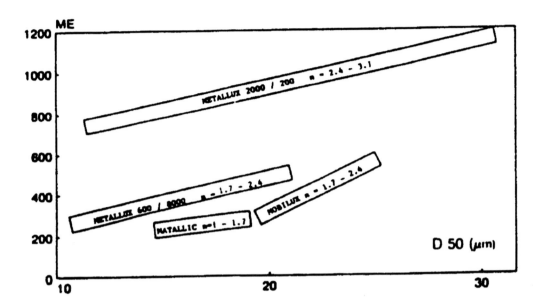

Figure 4. Optical properties of aluminum pigments as a function of particles size distribution and flake type. ME - metallic effect; s - coefficient of particle size distribution (1 - wide, 3 - narrow).

Table 1 and Figures 3 and 4 illustrate the relation between particle size distribution, particle size, their shape and the metallic effect.

The finer aluminum types appear darker, and particularly in the fine range of 3-10 μm, have a graying effect in combination with not very bright colorants.

To come back to the question of hiding. The reason why better hiding is achieved with smaller metallic pigments is because they form more of a multilayered barrier. This barrier allows less light to penetrate than when formed by large particles.

Aluminum pigments with a wider particle size distribution (industrial grade) contain mostly enough fine particles to hide well even if the average particle size is coarser. The particle size of metallic pigments is categorized as follows:

fine	6-15 microns
medium	16-22 microns
coarse	23-26 microns
extra coarse	27-32 microns

Comparing the particle size of fine metallic to spherical pigments explains why finer aluminum pigments hide better. Outside of the fact that they are flat, fine metallic pigments

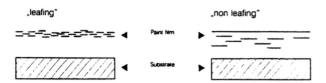

Figure 5. Comparison of leafing and non-leafing pigments.

are at least 10-20 times larger than well dispersed organic pigments. The particle size of spherical pigments is below 1 μm. At that size their ability to scatter light is substantially reduced.

Aluminum pigments are offered in a leafing and a non-leafing version. In the leafing version, the pigments are treated with stearic acid which causes the flakes to float to the surface in a liquid system like paint. The non-leafing aluminum pigments are surface treated with oleic acid. They do not float to the surface, but are distributed more throughout the system. They also display some orientation (Figure 5).

CONCEPT OF LEAFING VS. NON-LEAFING

For the leafing effect to take place, the system into which the metallic pigments are incorporated must have initially a low viscosity. Since for most plastic applications this is not the case, in general only non-leafing pigments are being used. Due to the high viscosity of most thermoplastic resins, non-leafing pigments display little orientation in these systems, but it was found that the lower the viscosity of the resin during processing, the brighter the resulting metallic shades due to improved orientation.

Aluminum pigments are not overly resistant to acids, alkali, chlorine, etc. For most plastic applications, this is of minor importance because these types of chemicals are not present during processing. Once the metallic pigments are imbedded into the plastic, the resins protect their surface and these chemicals have little effect.

PVC, on the other hand, can release chlorine during the processing. To avoid any chemical reaction with the aluminum pigments, encapsulated versions are available. One product offered is silica encapsulated (PCR) and an other is coated with a polymer (PCA). The polymer coated pigments are more brilliant and are slightly easier to incorporate into plastics. The encapsulated metallic pigments have the added advantage that they require more energy to be ignited. In the case of PCR, 12 times more energy is required. However, all aluminum powders must be handled with care. More about them later.

Aluminum pigments are offered as powder, pastes, and granules (concentrates). Particularly in recent years, the paste and recently the granules have been preferred over the powder for reasons of cleanliness and safety. The pastes are offered with different carriers. For polyolefins and styrenics, mineral oil based pastes have been favored. While for PVC, plasticizer based products are chosen.

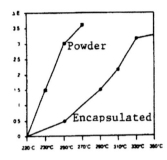

Figure 6. Heat stability of bronze pigments.

Granules are aluminum concentrates at a very high pigment loading, which have only been introduced recently. They offer the advantage of being predispersed and combine ease of handling with a much reduced chance of damaging the aluminum particle during the dispersing process.

GOLD BRONZE PIGMENTS

Bronze pigments are alloys of copper and zinc. The different ratios at which they are blended are responsible for the different shades of gold. The names of these gold shades have been standardized throughout the industry:

$$Cu:Zn = \quad \begin{matrix} 70:30 & \text{Richgold} \\ 85:15 & \text{RichPalegold} \\ 90:10 & \text{Palegold} \\ 100:00 & \text{Copper} \end{matrix}$$

Similar to aluminum pigments, bronze pigments are also offered in different particle sizes. Again the larger the particles, the brighter and more brilliant the metallic effect and the smaller the particle, the better the hiding. However, the amount of fines present, affect the brightness of the color less than is the case for aluminum pigments.

Contrary to aluminum pigments, bronze powders are sensitive to oxidation during storage and when exposed to heat. In both cases a color change is the result. To circumvent this problem, a silane coated version is available (Resist-Types). The coated bronze pigments are heat stable up to 450°F and have displayed no color change after three years of storage (Figure 6). Also bronze pigments are available as powders, pastes and possibly in the future also as granules (Table 2).

Table 2. Gold bronze pigments

Product	Pigment content, %	Carrier
Powder	100%	
Encapsulated pigments	100%	Silane (R-types)
Pastes	86%	Plasticizer, mineral oil, solvents
Pellets	95%	Pending

OPTICAL CHARACTERISTICS IN PLASTICS

The metallic effect is not only influenced by the optical characteristics of the metallic pigment, but also by the clarity of the plastic. Resins like PS, PC and PVC give the best effect because of their high degree of transparency. While polyolefins, ABS and other milky resins will offer less brilliant metallic appearance. For these resins, the metallic effect can be enhanced by raising the amount of metallic pigment from 0.5-1% to 2-3% aluminum pigment.

When using bronze pigment 1-3% is necessary. The reason why more bronze than aluminum pigment is required is due to the fact that its specific gravity is higher. In the average 2.7 times more bronze than aluminum pigment is needed.

The metallic effect is dependent on the incoming light being reflected to the highest possible degree. Not only a turbid resin, but also additives that are not completely translucent can interfere with the reflection of the light, i.e., the metallic effect.

Hiding pigments as TiO_2, most inorganic colorants and some organic pigments can have a negative effect on the metallic appearance. It is therefore important to chose colorants which are as transparent as possible for optimum brilliance, e.g., transparent iron oxides or Pigment Yellow 83. The least amount of light scattering and best colored metallic effect can be achieved by using dyes instead of pigments. However, dyes will only solubilize in resins that offer a certain polarity, such as PS, PC, ABS and PMMA. In polyolefins they are mostly insoluble.

By combining metallic pigments with pearlescent pigments, unique optical effects can be achieved. Small additions of metallic pigments to material colored with pearlescent pigments will improve the hiding substantially.

PRE-BLENDING

Contrary to most other pigments, metallic pigments are very easy to disperse. The reason is that they are coated during the manufacturing process with either stearic acid for leafing or oleic acid for non-leafing types. These additives provide a steric hindrance that prevent any van der Waal's forces to take effect and very little energy is required to disperse these types of pigments. On the contrary, if too much energy is used, i.e., the dispersing method is too harsh, the bronze or aluminum particles become damaged, bent and/or broken, resulting in a duller shade and in an extreme case, less hiding.

The best way of pre-blending metallic pigments with thermoplastic resins is by tumbling or through gentle stirring for not more than ten minutes. The action of a Henschel type mixer is too harsh and would damage the metal flake. It is therefore not recommended for blending aluminum pigments with plastic resins. In order not to expose the metal flakes to any severe shearing action that might be required for other ingredients, it is advisable to incorporate them at the end.

In general metallic pastes are preferred over powder to avoid dusting and the danger of a dust explosion when mixed with powdered resin. The carrier for the pastes can vary and is selected for its compatibility with the resin into which they are incorporated. For polyolefin, mineral oil is mainly used while for PVC, various plasticizers are preferred. The pigment concentration of the pastes range from 65-80% pigment.

Bronze pigments have a lower propensity to dust because of their higher specific gravity and are more frequently used as powder. Some aluminum pigments are also available as granules which are highly concentrated dispersions that are compatible with some of the major thermoplastic resins (Table 3).

If difficulties are encountered during incorporating the metallic pastes, they can be diluted at a ratio of 1:1 with, e.g., paraffin oil, lubricants, surfactants, or plasticizers. The well mixed blend is then incorporated into the plastic. When blending the plastic with the granules, the danger of damaging the flakes in the blending process has been substantially reduced because the metal particles are now protected by the carrier resin. A short blending time on the Henschel type mixer should in this case not harm the metallic flakes.

If the plastic is liquid as is the case of reactive polyester (UPE) or methyl methacrylate resins, the metal paste should first be diluted with a liquid like styrene in case of UPE at a ratio of 1:1. This homogeneous blend is then again diluted with a liquid (1:4= blend:liquid) before adding it to the final formulation.

Table 3. Aluminum pigments. Available technologies

Product	Pigment content, %	Carrier
Powder	100%	
Encapsulated pigment	100%	Silica (PCR-types), polymer (PCA-types)
Pastes	65-80%	Plasticizer, mineral oil, polyol, solvents
Pellets	80-95%	Polyolefin, proprietary blends

DISPERSION ON DIFFERENT EQUIPMENT

INJECTION MOLDING MACHINE AND EXTRUDER

In most cases the dispersing action achieved, be the screw of these type of machines, is sufficient to achieve a good distribution of the metal pigment. To improve the pigment dispersion, increasing the back pressure caused by the screen pack has been found effective.

TWO ROLL MILL AND CALENDERING

In general, metallic pigments disperse very easily on the two roll mill due to their strong friction, but care needs to be taken not to damage the metallic flakes. During the calendering process, any excess material or trim containing uncoated bronze pigment should not be recycled because the additional exposure to shear and elevated temperatures can cause a color change, fire, or explosion. Coated bronze pigments such as the Resist type overcame this problem.

DIFFERENT PLASTIC RESINS
PVC

Table 4. Encapsulated metallic pigments

Type of treatment	Applications	Performance characteristics
Aluminum pigments, silica encapsulated, (PCR-type)	PVC, powder coating, thermoplastic resins, thermosetting resins	Improved chemical resistance and outdoor durability
Aluminum pigments, polymer coated (PCA-type)	PVC, powder coating, thermoplastic resins, thermosetting resins	Improved chemical resistance and outdoor durability; better resin compatibility
Bronze pigments, silane coated (Resist type)	PVC, powder coating, thermoplastic resins, thermosetting resins	Improved heat and storage stability; better resin compatibility

As mentioned, in some cases hydrochloric acid is released during processing of PVC which could attack the metallic pigments if they are not encapsulated. It is, therefore, advisable to use the encapsulated PCR or PCA aluminum pigments or the Resist bronze pigments, which are offered as powder or plasticizer pastes (Table 4).

The encapsulated products have also the advantage that no traces of copper or zinc ions are introduced which can cause a catalytic degradation of the PVC. Bronze pigments, if used in plastisol, have a tendency to settle due to their high specific gravity. The addition of fumed silica should counteract the settling.

To stabilize PVC, Ba/Cd-stabilizer or organotin- stabilizer is recommended because of their good transparency. In addition they have proven to be excellent light stabilizer with little odor. Lead stabilizers introduce a certain haze which will negatively effect the metallic appearance desired.

Polyolefins

The metallic pigments are either incorporated as granules or as mineral oil pastes. Because the paraffin oil has a lubricating effect which can cause slippage at high concentrations, i.e., in case of master batches, a Banbury or a twin-screw extruder should be used for best results. Since the transparency of polyolefins increases with higher density, the best metallic effects are achieved with PP and also by increasing the pigment loading somewhat. For bronze pigments, only the encapsulated version is recommended to insure proper heat stability.

If fire retardant additives are being used based on halogens, they also can react with metallic pigments. To avoid a chemical reaction, either the encapsulated aluminum or the encapsulated bronze pigment is recommended. Also it is possible to add the metallic pigment as a masterbatch, thereby avoiding direct contact and a chemical reaction between the metal and the halogens.

Styrenics

Again pastes are preferred over powder. During the mixing process care should be taken that the sharp edges of the PS do not damage the flakes of the metallic pigments. The mixing time should therefore be as short as possible to avoid a change of color or brightness.

ABS or HIPS have a milky and sometimes slightly yellow appearance which require a higher pigment level of 2-3%. To obtain a homogenous blend at the higher pigmentation levels, it is recommended to introduce the metallic pigment as a masterbatch because larger amounts of aluminum pigment will not sufficiently adhere to the surface of the PS granules.

As mentioned, the surface of aluminum pigments is treated with fatty acids which can cause a rancid odor during processing. By adding an antioxidant, the problem can be overcome.

CONCLUSIONS

The most brilliant metallic affects are achievable by aluminum or bronze pigments with a large particle size. The brilliance can be increased further by selecting products with a particle size distribution that is as narrow as possible thereby reducing the amount of fines present. It is also important to allow the flake like metallic pigments to orient themselves parallel to the surface to optimize the light reflection. It was found that the lower the viscosity of the plastic resin during processing, the more the metal flakes will orient and the better the metallic affect. Clear resins like PS or PVC allow to achieve a brighter metallic shade compared to resins which are turbid like polyolefins. Independent of the resin the brightness of metallic pigments decreases with decreasing particle size, while their ability to hide increases.

Encapsulated metallic pigments were developed to improve their chemical resistance allowing for instance their application in PVC. By encapsulating the bronze pigments their heat resistance and storage stability was also substantially improved permitting them to be applied in most plastic resins without darkening.

When incorporating metallic pigments into plastics care needs to be taken as not to damage the flake during the dispersing process which would result in reduced brilliance. Since aluminum and bronze pigments disperse very easy, this represents no problem.

REFERENCES

1   G. Sommer, **Optical characterisation of metallic print**, *Eckart-Werke,* Eckart Metallpigment fuer die graphische Industrie.
2   R. Besold, **Qualitiy control test methods**, *Echart-Werke*, Alu-Flake Schriftenreihe 2
3   E. Roth, **Dispersing instructions for Stapa Metallic, Mobilux and Metallux pigments**, *Eckart-Werke* Alu-Flake Schriftenreihe 2.
4   L Gall, **Farbmetrik auf dem Pigementgebiet**, *BASF,* 1970, S 399 Publikation.

Metallic Looking Plastics. With New Silver and Colored Aluminum Pigments

Hans-Henning Bunge
Eckart America L.P., USA

ABSTRACT

The application of aluminum pigments in plastics presented a problem in the past because of the flow lines which they caused in injection molded parts. By using large size aluminum pigments, with an average particle size of 60 µm to 330 µm and larger, it is possible to avoid these flow lines and produce metallic looking plastic parts. This concept was taken one step further by depositing colorants on these large size aluminum pigments, thereby creating a blue, green, and golden metallic looking colorant that can be used in plastics without flow lines. These pigments offer exceptional styling effects by themselves and in combination with other colorants including bronze pigments.

INTRODUCTION

Aluminum pigments have not been favored in the past for injection molded parts for more than one reason. If they were used as powders there was always a concern that they could present a hazard. This was widely overcome by using these pigments pasted in polyethylene wax or in a predispersed form as concentrates provided by their supplier. The second reason was that aluminum pigments cause flow lines in injection molded parts which in most cases can not be tolerated.

APPLICATION OF ALUMINUM PIGMENTS IN PLASTICS

Aluminum pigments did find wide application in plastic films for garbage bags, canvases, as heat reflectors and for agricultural applications. Because plastic films are inherently thin, aluminum pigments with only a relatively small average particle size in the range of 10 µm and below are applied.

Aluminum particles of this size display two main properties, excellent hiding but decreasing brightness with decreasing particle size. Excellent hiding, because contrary to regular pigments, the hiding of aluminum pigments increases as the particle size decreases due to their lamellar structure. But the smaller the aluminum particles the darker their appearance because of the increasing amount of light that is scattered. On the other hand, the larger the aluminum flake the more light is reflected and the brighter and the more metallic like its appearance. Again this brightness is reduced by the amount of smaller particles or fines that are present.

AVOIDING FLOW LINES

For plastics, large particles have an additional advantage. Aluminum flakes starting at a particle size of around 125 μm to 700 μm and even above cause in injection molded parts no flow lines or in the 60 μm range strongly diminished flow lines.

The following additional steps[1] can be taken to insure the elimination of flow lines and thereby optimizing the optical appearance of injection molded parts prepared with such aluminum pigments:

- optimum concentration of the aluminum pigment
- resins with a high viscosity (low MFI) reduce flow lines, because the aluminum flakes will orient less
- a strong turbulence in the mold, created by large diameter of the gates and a high injection velocity
- when designing the mold the wall thickness of the molded part should be optimized as well as the locations of the gates
- for large parts cascading gates are preferred.

METALLIC LOOKING PLASTIC PARTS

Because of the high reflectivity of these large size aluminum pigments they can provide metal like appearance of injection molded parts. These type of aluminum pigments are available as powders, pastes in mineral oil, or predispersed as pellets. The pellets which are the most widely used version have a pigment concentration ranging from 80-90%. Some of these pellets are dispersed in polyethylene wax only, restricting their use primarily to polyolefin resins. Others contain only a small percentage of polyolefin wax in combination with other additives widening their application to most commercially available resins.

The advantages of the predispersed version are that they are much easier to handle. Also they have been dispersed on equipment that is designed to do as little as possible damage to the flakes during the extrusion process because aluminum flakes that are bend during processing are less bright.

To make these flakes tougher and more resistant to bending they are offered by one supplier at a thickness of 8-10 μm compared to 2 μm of regular aluminum pigments. These thicker flakes are available at particle sizes of 125 μm, 225 μm and 330 μm. The larger the particle the less their ability to hide. Therefore to achieve a convincing metallic appearance the pigment loading has to be increased in relation to the particle size. In some cases the best effect is obtained at a 10% aluminum pigment concentration particularly if the average particle size is 330 μm and higher.

EFFECT ON PHYSICAL PROPERTIES

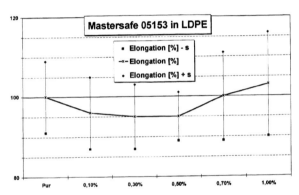

Figure 1. Elongation vs. concentration of Mastersafe 05153 having average particle size of 5 μm.

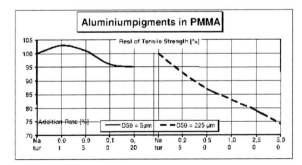

Figure 2. Tensile strength vs. pigment concentration.

With increasing pigment concentration some of the physical properties start to suffer. As is evident from these measurements the influence of aluminum pigments on the physical properties depends on the loading as well as on their particle size. Aluminum pigments with a particle size of 5 μm which are mainly applied in films have little or no effect on the film properties. While aluminum with a particle size of which we are talking about here, i.e., 60 μm and larger reduce the tensile strength notably.

These properties will differ in chemically different resins but even chemically similar resins by different supplier can show different performance characteristics. However, in many cases the lowering of the physical properties can be tolerated depending on the end use of the injection molded part. It should be noted based on customer reports that the physical properties of fiber reinforced polyamide parts are not effected by such high aluminum pigment loading.

APPLICATIONS

Besides handles, toys, knobs and other plastic objects where a metallic appearance are desirable these type of aluminum pigments are widely applied in automotive parts under the hood.

These automotive parts are mostly made from reinforced polyamide resins and are presently mainly colored black. However, the automotive industry is making a strong effort to give these parts a metallic appearance. Companies as VW and BMW are already equipping there cars with metallic looking engine covers and others are also in the process of converting from black to a silvery look.

COLORED ALUMINUM PIGMENTS

This concept of obtaining injection molded parts without flow lines was carried one step further by creating colored aluminum pigments at a particle size of 125 μm. At the present time three colors are available gold, blue and green. The colored aluminum pigments are manufactured by depositing organic pigments on the aluminum which is then encapsulated by a silica coating.

To insure excellent heat stability and good out door durability of these pigments only colorants were selected which met these requirements like phthalo-blue and green. This gives them a distinct advantage over cut aluminum foil which has mostly borderline heat stability and is coated with a colored ink film. By adding a pigment layer plus a silica layer during the production of these colored aluminum pigments their thickness increases to 12-15 μm making them even less fragile during the dispersing and injection molding process thereby insuring a brilliant, specular effect of the injection molded part.

STYLING POSSIBILITIES

Due to their own color these aluminum pigments offer unique styling possibilities. The best effects are achieved in combination with dark colorants including black at a colored aluminum pigment level of 1.5%. In general, it is recommended to combine aluminum pigments with transparent colorants because opaque pigments will hide the metallic effect.

However combinations of the colored aluminum pigments with different bronze pigments offer very unusual rich looking shades. For these combinations the gold colored aluminum is less suited than blue and green.

PIGMENT CONCENTRATION OF COLORED ALUMINUM

When the object is to create a metallic looking plastic part, the function of the aluminum pigment is to form a uniform layer within the molded product for which a high pigment loading up to 10% can be necessary. With colored aluminum, the object is to create unique color effects in combination with other mostly transparent colorants. To avoid that the other colors are blocked out by the aluminum pigment. The loading of the metallic colorant should be fairly low ranging from 0.2% to 1.5%.

CONCLUSIONS

By offering silver and colored metallic pigments, which avoid flow lines of injection molded plastic parts due to their large particle size, totally new possibilities have been opened for achieving brilliant, metallic like appearance and new, unique color effects of plastic products.

REFERENCES

1   B. Klein/H-H.Bunge, Cost Reduction by Metallic Pigments, *Kunststoffe*, 9/96.

Ultramarine Blue, an Old pigment, a New Process

Thierry Guilmin
Prayon Pigments sa.

ABSTRACT

Ultramarine Blue is an inorganic pigment used for centuries all over the world. This Sodium Alumino Sulfosilicate is known as Lapis Lazuli in its natural form. In 1826, J. B. Guimet invented the synthetic Ultramarine Blue produced through a batch process. It is only in 1993 that Prayon Rupel invented the continuous production process for this pigment. This process has been evaluated and developed for the last 4 years on a pilot plant. Today, the industrial plant is running under the same conditions as the pilot plant. This evolution in production has to bring at least some quality improvement in pigments over the batch process and preferably some specific advantages for the end-users. Some of these studies will be presented in this paper.

INTRODUCTION

The end-user of a pigment needs specific characteristics from which the most important are:
1. The color (L, a, b or L, c, H)
2. The reproducibility and the consistency
3. The predictability in his formulations

Other aspects such as tinting strength, dispersability, chemical resistance ... are also of importance but will not be discussed this time.

THE COLOR

When you buy or use a pigment, you use it for its color. This obvious statement is difficult to control in practice as everyone looks at color with his own eyes. Fortunately, the spectrometers give CIELAB values which are accepted by everyone. This system is valid for comparison only if a pigment is analyzed in the same matrix or resin.

Ultramarine blue pigments give similar L, a, b values when they are compared in the same matrix and the first point was to confirm that the batch process and continuous process give similar values. Table 1 illustrates the fact that the production process gives similar values in both cases.

Table 1. Color data for different processes

Product	Production	Shade	L value	a value	b value
R950201C	batch	red	74.4	-2.9	-25.7
R960118P	batch	red	74.6	-3.2	-27.7
L960223	continuous	red	73.8	-3.1	-26.6
R940222B	batch	green	72.8	-4.7	-28.7
R941110B	batch	green	73.0	-4.7	-29.1
L960613Z	continuous	green	72.6	-4.7	-28.7

These measurements were done according to the DIN55907 method in alkyd film, dilution in oil, 20% pigment and 80% TiO_2. These lots come from various producers and confirm the capability of the continuous production process to make similar pigments.

THE REPRODUCIBILITY AND CONSISTENCY

The next step in the validation of the production method is the reproducibility of the lots produced. This had to be compared to the reproducibility of the batch process. The best way to control this is to check the color parameters, and more precisely on the Δ values of these color parameters. The references taken were the medium sample values. The samples are supposed to cover 2 years of production from specific Ultramarine Blue pigments.

Table 2 illustrates this survey. It appears that both green and red shade Ultramarine Blue have a ΔE </= 0.7, which confirms the consistency of production. The progress done during the last years in the batch process is shown by the values of ΔE inferior to 1. The difference is that in the case of the continuous process, this consistency is easier to get. In final formulations, it is important to get a ΔE for each involved pigment as small as possible and that is why a continuous process brings improvement.

Time consuming evaluations of each incoming lot of material is the next step to improve. If pigment producers can offer narrow specifications, a simple input in spectrometer for color matching is the next step.

Table 2. Reproducibility and consistency

Product	Shade	L*	a*	b*	#	ΔL	Δa	Δb	ΔE
R970514	red	74.8	-2.5	-26.3	30	0.3	0.6	0.8	1.0
R970325	red	74.8	-3.1	-27.4	5	0.9	0.2	1.2	1.5
L951213	**red**	**73.9**	**-3.1**	**-26.7**	**14**	**0.4**	**0.3**	**0.5**	**0.7**
R970214	green	73.0	-4.1	-30.1	15	0.3	0.4	0.6	0.8
R970513	green	73.0	-4.0	-29.2	12	0.3	0.4	0.7	0.9
L950516	**green**	**73.1**	**-4.8**	**-28.6**	**11**	**0.4**	**0.3**	**0.4**	**0.7**

#=number of samples

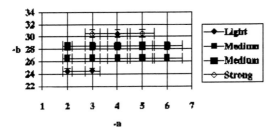

Figure 1. Ultramarine blue color scale.

PREDICTABILITY IN FORMULATIONS

Color matching is a difficult matter. Introducing a direct link between product data sheet specifications and supplied material is the best way to help people in solving the color matching problem.

This statement was taken into account in the continuous production process and Figure 1 illustrates this approach. As you can notice, by building up this product range, it becomes possible to cover most of the color spectrum of Ultramarine Blue pigments.

When a production unit can remain in these areas for each grade, the color matching is easier and these values can even be introduced in the spectrometers, in order to work directly with standard pigments. The L values are not in this projection as they are more dependant of the C black or TiO_2 content in colored formulations. This is based on the DIN 55907 method with alkyd resins. The incorporation in various matrixes such as polymers, acrylics, nitrocellulose will slightly modify these values but they will always be modified the same way, so in delta E, the impact is not important.

CONCLUSIONS

A continuous production process of Ultramarine Blue pigment answers the demand on color as well as existing batch production process, improves on consistency and helps in final formulation.

These are the first parameters to satisfy color and other factors (tinting strength, dispersability, thermal stability, weatherability..) that will influence the final formulations but these are still under evaluation.

Predicting Maximum Field Service Temperatures From Solar Reflectance. Measurements of Vinyl

Henry K. Hardcastle III
Dayton Technologies

ABSTRACT

Vinyl products continue penetrating Western US markets. Vinyl products may show unacceptable heat distortion when installed in Western environments even after demonstrating a long tradition of acceptable heat build performance in Eastern US environments. This paper presents a methodology for predicting maximum field service temperatures from solar reflectance measurements. Solar reflectance data (ASTM E-903 and E-892), field measurement data and a predictive model for a variety of vinyl systems are shown. This methodology may be used in addition to ASTM D-4803 and is not limited to vinyl materials.

INTRODUCTION

A number of vinyl building product manufacturers are familiar with The Standard Test Method for Predicting Heat Buildup in PVC Building Products according to ASTM D 4803 which utilizes an insulated box to house a specimen irradiated by an IR heat lamp. Many vinyl producers may not be familiar with the basis of this test or the direct measurements that can be made to predict the propensity for heat buildup.[1] Recent failures of rigid vinyl materials due to heat buildup and heat distortions have been observed even though ASTM D-4803 analysis indicate acceptable performance. These materials have also displayed satisfactory heat buildup performance in historical markets. Sales and subsequent failures of these products in newer Western US markets imply an environmental constraint not found in traditional eastern geography's and a possible limitation to the D-4803 method. Failures that initiated this study have been focused around areas with higher solar irradiance in the Southwestern US.

STATEMENT OF THEORY AND DEFINITIONS

THE SOLAR SPECTRUM

The solar spectrum is a depiction of the energy from the sun that irradiates a material. Due to filtering effects of the atmosphere more than 98% of the sun's energy that strike the earth's surface are between 300 and 2500 nm. The irradiant energy at any particular wave band within this spectrum is highly dependent on the amount and quality of atmosphere the energy travels through before striking the material.

- There are several different agreed upon solar spectrums
- One of the major differences is the amount of atmosphere the energy must travel through
- Another difference is the amount of direct vs. diffuse light irradiating the surface
- Three major solar spectrums defined are Air Mass 1.5 Direct, Air Mass 1.5 Global and Air Mass 0 as shown in Figure 1.
- There may be other sources of irradiance besides the sun contributing to heat build including; shingles that are reflecting or re-radiating at long wavelengths, low E glass, barbeque grills, pool decks and other good absorbers, emitters or reflectors of solar energy. Often these features may concentrate solar energy or re-radiate absorbed solar energy at longer wavelengths and contribute additional energy for heat buildup.

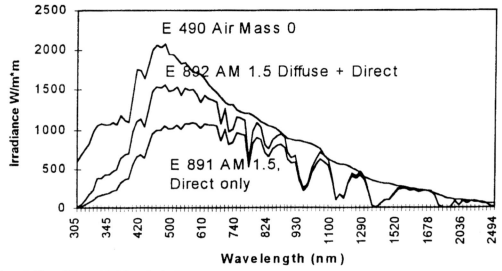

Figure 1. Three different ASTM standard solar spectrums.

VINYL OPTICAL PROPERTIES

Optical properties can be characterized using the relationship:

$$1 = \rho + \tau + \alpha \qquad\qquad [1]$$

Where ρ represents the solar energy reflected from the material, τ represents the solar energy transmitted through the material and α represents the solar energy absorbed by the material. The relationship simply states that the total irradiance striking a material will either be reflected off the material, transmitted through the material, or absorbed by the material. It is the absorbed solar energy that is available for heat buildup.[2]

- The relationship becomes even more simple if the material is opaque ($\tau = 0$)
- It is important to consider a materials optical properties throughout the entire solar spectrum (approximately 300 to 2500 nm) rather than just the visible spectrum or just the IR spectrum since about half of the solar energy is composed of wavelengths less than 780 nm and half the solar energy lies above 780 nm
- Some materials that have low absorptance in the visible portion of the solar spectrum may have high absorptance in the IR region. Pigment manufacturer's take advantage of this fact and produce many products often referred to as "IR reflective pigments" that appear dark in visible light but are highly reflective in the IR and therefore remain cooler than similar colors made with traditional pigments

DESCRIPTION OF EQUIPMENT AND PROCESSES

MEASUREMENT OF OPTICAL PROPERTIES

Measurement of reflectance and transmittance optical properties is easily accomplished using modern commercially available spectrophotometers.

It is important that the spectrophotometer has the ability to scan the majority of the solar spectrum from approximately 300 to 2500 nm.

The geometry of the measurement, incident and reflected angle of spectrophotometer beams, reference beams and use of integrating spheres are important considerations of these measurements especially when comparing optical properties measured using different configurations or instruments.

Measurement geometry and front end optical designs are well documented in ASTM E903 for these measurements.

The initial result of these optical properties measurements is typically a spectral reflectance or transmittance curve showing the %ρ or %τ at each wavelength as a graph as shown in Figure 2.

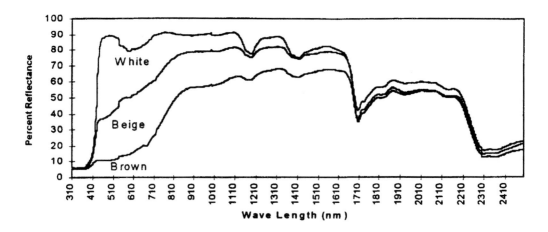

Figure 2. Spectral reflectance curves for 3 colors of rigid vinyl.

INTEGRATION OF OPTICAL PROPERTIES TO THE SOLAR SPECTRUM

Once a measurement of the percent reflectance and percent transmittance of the material at each wavelength is obtained through out the solar spectral region (300 - 2500 nm) the optical properties of the material may be related to the sun's irradiance by integration.

Integration is a mathematical weighting process that takes into account both the sun's irradiance and the material's reflectance at each wavelength from 300 to 2500 nm. Integration weights regions of the material's optical properties spectrum according to the energy output from the sun in those regions.

Once the sun's irradiance and material's optical properties are integrated at each wavelength, the total of reflected solar energy may be summed resulting in a single number denoted as "total percent solar reflectance" for the air mass used. Percent solar absorptance is then calculated:

$$\alpha = 1 - (\rho + \tau) \qquad\qquad [2]$$

It is the value of percent solar reflectance and the calculated percent solar absorptance that is powerful in predicting a materials propensity for heat build.

For opaque materials such as a rigid vinyl, colors with high solar reflectance will remain cooler than colors with low solar reflectance under the same environmental conditions.

For materials with the same emmittance characteristics, materials with higher solar absorptance will have a greater propensity for heat build. Materials with lower solar absorptance should remain cooler for similar materials under the same solar and ambient conditions.

APPLICATION OF EQUIPMENT AND PROCESSES

There appear to be 4 main steps to using the solar spectrum, optical property measurements and solar integration;
1) Define the temperature failure criteria for the material.
2) Obtain empirical heat build up data for a number of material colors in worst case environments.
3) Measure the optical properties of the material colors and plot correlation regression between solar absorptance and worst case empirical heat build data noting where the regression line crosses the failure criteria.
4) Consider the risks involved with selling products which measure above the critical solar absorptance characterized in the previous step. An example will demonstrate use of these four steps.

EXAMPLE OF METHODOLOGY

A producer offers a variety of different colors in the same PVC base. Colors are formulated by altering the pigments and TiO_2 content. In this example, the producer has no prior knowledge of field performance but wants to determine the maximum solar absorptance he can design and still have acceptable heat buildup performance.

1) Define the temperature failure criteria for the material. The producer determines experimentally the maximum service temperature his material can achieve and still provide acceptable performance. The producer determines the heat deflection temperature (ASTM D 648), Vicat Softening Temperature (ASTM D 1525), Coefficient of Thermal Expansion (ASTM D 696) or other appropriate quantitative measure of material's performance under heat. The producer then adds a suitable safety factor to the temperature determined to cause failure.

2) Obtain empirical heat build up data for a number of material colors in worst case environments. The producer obtains a number of samples of different colors of his material and exposes them to the worst case environment in his intended market. This environment should have the highest solar irradiance and warmest temperatures the product may be subjected to while in service. The samples should be oriented for exposure resulting in the maximum heat build; oriented normal to sun, protected from breezes and insulated from convective and conductive cooling as much as appropriate for the product. Consideration should also be given to reflective surfaces and other heat sources the product may encounter in the field. The producer then carefully measures the temperatures the selected samples reach under these worst

case conditions using thermocouples, pyrometers or other suitable temperature measuring and data logging instrumentation. The temperature measurements are made simultaneously for all specimens to block differences in environmental variables as shown in Figure 3.

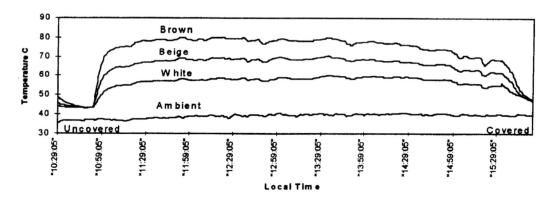

Figure 3. Heat buildup near summer solstice in Phoenix, AZ for 3 colors of rigid vinyl.

3) Measure the optical properties of the material colors and plot correlation regression between solar absorptance and worst case empirical heat build data noting where the regression line crosses the failure criteria. The producer then measures the solar optical properties of the samples used to obtain the worst case heat build temperatures and calculates solar absorptance. An x-y scatter plot is then constructed with maximum temperature on the ordinate and solar absorptance on the abscissa. The regression line is fitted to the data. The temperature failure criteria from step 2 is marked on the ordinate scale and a line is extended to intersect with the regression line as shown in Figure 4. The point of intersection with the regression line is then extended down to the abscissa and solar absorptance indicated becomes the design criteria for new products.

4) Consider the risks involved with selling products which measure above the critical solar absorptance characterized in the previous step. Products made with higher solar absorptances have a higher risk of exceeding the defined temperature failure criteria determined in step 1. Again, the producer may choose to utilize IR reflecting pigments to produce dark colors or decide the cost and risk outweigh the revenues from color offerings with higher solar absorptance.

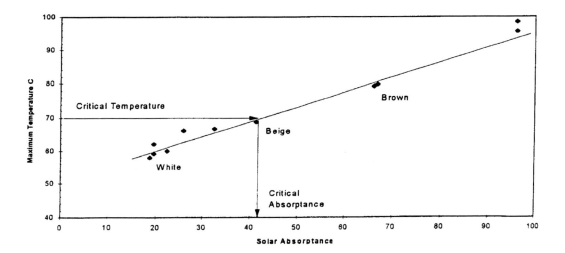

Figure 4. Maximum temperature vs. solar absorptance scatter plot denoting critical absorptance for a rigid vinyl system.

PRESENTATION OF DATA AND RESULTS

ACTUAL CASE STUDY DATA

The company had been selling traditional colors of window lineals since the mid 1980's with acceptable heat build performance. Colors sold included white, beige and brown in standard rigid vinyl formulations. The company utilized ASTM D 4803 to evaluate heat buildup before a new color's introduction. In the past several years, however, several things happened that increased risk of heat related failures;

1) Markets moved westward from northeastern and southeastern and central US areas to southwestern US markets such as Denver, Las Vegas, Phoenix, Southern California, etc.

2) Customers began demanding new custom colors including very dark colors.

3) The company began experimenting with new formulations including different types and amounts of lubricants, stabilizers, impact modifiers, and colorant vehicles.

4) Heat related problems became a focus of discussions among building product producers and suppliers.

The challenge for the R&D effort was to develop a methodology to predict propensity for heat buildup for new experimental formulations in addition to the D4803 method. The new method needed to be empirically based and applicable to the data base of performance already available (e.g. customer complaints and historical product offerings). Finally, this method needed to provide decision makers with a clear indication of new products performance before release to the markets.

1) Define the temperature failure criteria for the material. The experimental formulas were blended and extruded. The extruded products were measured for heat deflection temperature using ASTM D 648 as a guideline. Multiple measurements at various heating rates were conducted. An appropriate engineering safety factor was applied to the data. A critical temperature failure criteria was defined as 70°C for these particular experimental formulas. 70°C was considered the maximum sustained temperature the extrusions could withstand and still provide acceptable engineering performance.

2) Obtain empirical heat build up data for a number of material colors in worst case environments. A collection of 11 specimens representing the range of current product offerings and R&D efforts was selected. The materials were mounted in a single standard frame, side by side. The specimens were similar in thickness and dimension. The frame and specimens were mounted over standard building insulation to prevent back side cooling and surrounded by wind baffles to reduce cooling due to breezes. Thermocouples attached specimens to a simple data logger. The specimens were exposed directly to sun at Phoenix, AZ near summer solstice 1997 at near normal angles. Measurements were taken continuously for several days. The maximum temperature achieved by all specimens at the same time was recorded. An example of this data is shown in Figure 3. These values were then described as the best estimate of heat buildup for the colors in a worst case environment.

3) Measure the optical properties of the material colors and plot correlation regression between solar absorptance and worst case empirical heat build data noting where the regression line crosses the failure criteria. Each of the materials was then measured using ASTM E 903 and integrated using ASTM E 892. Each material was opaque. The percent solar absorptance was calculated for each material. The solar absorptance vs. maximum heat build were plotted in x-y scatter plot format and fitted with a regression line. The temperature failure criteria was noted on the temperature scale and extended to the regression line. The point of intersection denoted the maximum % solar absorptance that could be achieved by the system and still provide acceptable heat buildup performance as shown in Figure 4. For this formulation, the maximum solar absorptance should not exceed 40%α critical value.

4) Consider the risks involved with selling products which measure below the critical solar reflectance characterized in the previous step. The critical solar absorptance value of 40% became a clear design criteria for current and new color product offerings in this system.

INTERPRETATION OF DATA

The empirically derived maximum temperature vs. solar absorptance regression shown in Figure 4 becomes an important tool for new product designers using this vinyl system. Different colors produced in this formulation can be identified on the regression by simply measuring their solar reflectance and calculating their solar absorptance value. Once a custom color is matched, a sample is immediately submitted for solar reflectance measurements. If a pigmentation system used to achieve a custom color results in solar absorptance values above the critical value, decision makers know the probability of heat related complaints will increase in severe environments.

SUMMARY AND CONCLUSIONS

Use of empirically derived heat buildup data and optical properties measurements can significantly improve a producer's ability to predict maximum field service temperatures of vinyl materials. Use of empirical field methods described here in addition to laboratory tests can identify robust design criteria, enhance a product's service performance and ultimately contribute to customer satisfaction.

ACKNOWLEDGMENT

The Author would like to acknowledge Dayton Technologies for permission to publish this work.

REFERENCES

1   E. B. Rabinovitch, *et al. J. Vinyl Tech.*, **5**, No. 3 (1983).
2   J. A. Duffie, W. A. Beckman, **Solar Engineering of Thermal Processes**, *John Wiley and Sons*, 1980, p. 144-154.

Reactive Trapping of 3,3'-Dichlorobenzidine Decomposition Products in Polyethylene-Based Diarylide Pigment Concentrates

William Anjowski
Colortech Inc., Brampton, ON L6T 3V1, Canada
Christopher J. B. Dobbin
Industrial Research+Development Institute, Midland, ON L4R 4L3, Canada

ABSTRACT

Concerns over the thermal decomposition products of diarylide pigments in polyethylene matrices have severely limited the use of this versatile and cost-effective pigment family in many colorant applications. A strategy for the reactive trapping of 3,3'-Dichlorobenzidine, a potential human carcinogen formed during the high temperature processing of polyethylene concentrates, is discussed. Chemical trapping tests made using maleic anhydride modified polymer additives showed favorable reactivity towards 3,3'-DCB in model systems. The results of laboratory screening trials with Pigment Yellow 13 and Pigment Yellow 83 are also reported. The apparent complexity of the diarylide pigment decomposition reaction in LLDPE at typical processing temperatures ($>200^{\circ}$C) made isolation, analysis and quantification of residual 3,3'-DCB levels extremely difficult.

INTRODUCTION

The last ten years have seen a steady decline in the use of lead and cadmium-based pigments in the coloration of polyethylene and polypropylene packaging materials and durable goods. Largely as a result of environmental concerns, these relatively inexpensive, highly opaque, lightfast pigments have been replaced with more expensive organic alternatives.

In the yellow-red region of the visible spectrum, traditional lead pigments such as chrome yellow and moly orange have been replaced with a variety of organic alternatives. In general, the replacement pigments are;

a   much more expensive,
b   much less strongly colored,
c   less opaque
d   possess poorer lightfastness.

Until 1990, certain members of the diarylide pigment family had found widespread use in polyolefin resin concentrates as cost-effective alternatives for lead and cadmium pigment replacement. In early 1990, Hoechst AG, Germany, a major manufacturer of diarylide pigments, issued a notification to its subsidiary companies advising them of laboratory research concerning the high temperature use of diarylide pigments in polymers.[1]

Subsequent Hoechst research[2] indicated that thermal decomposition of diarylide pigments could occur in polyolefin blends at temperatures above 200°C (392°F). Analysis showed that these pigments could undergo a cleavage reaction to form trace quantities of colored mono-azo dyes, which in turn could decompose further to form other aromatic amine derivatives. The amount and type of degradation product depended on the diarylide pigment(s) involved, melt temperature, formulation, processing conditions and dwell time. As conditions became more severe (i.e., temperatures of 240- 300°C or 464-572°F and dwell times of ten minutes or more), trace quantities of 3,3'-dichlorobenzidine (3,3'-DCB) were detected.

Because 3,3'-DCB is a known animal carcinogen and suspected human carcinogen,[3,4] Hoechst recommended that all applications involving the use of diarylide and pyrazolone pigments in polyolefin substrates at temperatures greater than 200°C be halted immediately pending the results of further testing. Virtually all other manufacturers of diarylide pigments echoed this recommendation within days of the Hoechst announcement.[5,6]

In general, two diarylide yellow grades (C.I. Pigment Yellow 83 and C.I. Pigment Yellow 13) had found widespread use in North American concentrate markets. Historically, Colortech had restricted its formulation work to the less expensive (and less thermally stable) PY13. Colortech had approximately 98 color formulations on file containing PY13 at levels from 0.5-15% at the time of the Hoechst announcement. In June 1990, Colortech declared a moratorium on the use of diarylide pigments, pending further evaluations and information. In the interim, little or no regulatory activity, coupled with a return to diarylide pigment use by competitive concentrate compounders has prompted Colortech to re-evaluate this situation.

THEORY

Az, Dewald and Schnaitmann[2] identified 3,3'-DCB as one of the primary products of thermal decomposition when polyolefin resins containing PY13 or PY83 pigments are subjected to processing temperatures above 200°C. The thermal decomposition mechanism is thought to proceed *via* bond scission reactions at the two azo linkages to produce non-azo derivatives

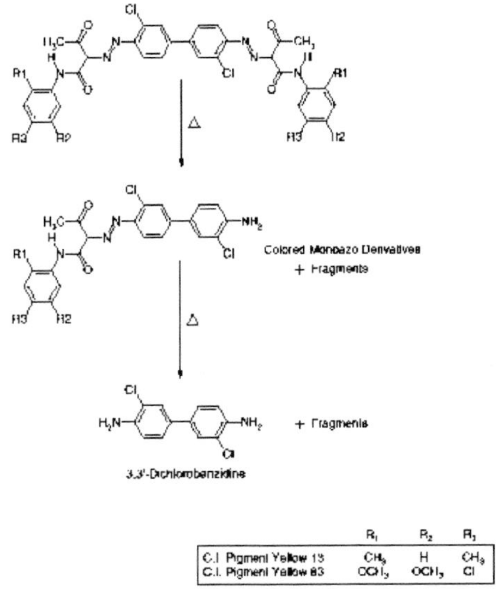

Figure 1A. Proposed reaction mechanism for diarylide pigment decomposition.[2]

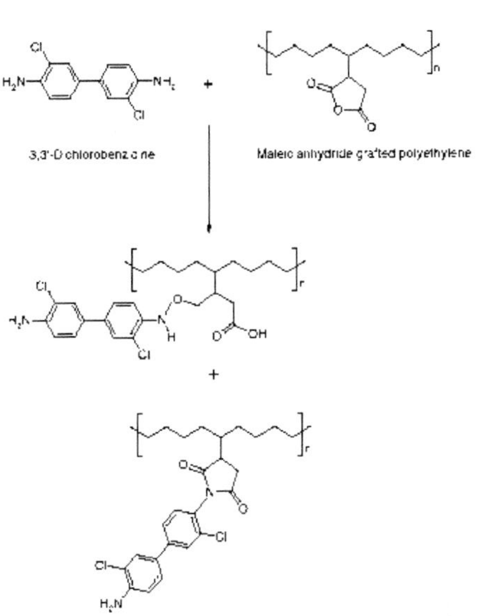

Figure 1B. Proposed trapping reaction for amine-based decomposition products.

that can subsequently decompose to 3,3'-dichlorobenzidine and a number of related amine functionalized fragments (Figure 1A).

Since the mono-azo intermediates and 3,3'-DCB all contain reactive amine functionalities, decomposition products could be trapped using chemically reactive polyolefin resins added to the color concentrate formulation. In principle, commercially available maleic anhydride-modified polyethylene resins and waxes possess the necessary functionality to react with amine compounds as they are produced. Potentially harmful 3,3'-DCB decomposition products would then be immobilized as pendant groups on the polymer backbone in the form of imide or amide/carboxylic acid pairs (Figure 1B).

Calculations suggest that even at low addition levels (i.e., 2-5% w/w in the concentrate), a maleated polyolefin incorporating approximately 1% grafted maleic anhydride would provide a large excess of reactive anhydride functionality in the concentrate. The effect of anhydride-grafted polymer addition in a typical dilute system (i.e., the concentrate "letdown") is the focus of the proposed work. It was hoped that this trapping strategy would provide a means of reducing "free" 3,3'-DCB levels and extending the safe use of diarylide pigments in polyolefin resins at moderate to high processing temperatures (225- 250°C).

EXPERIMENTAL

MATERIALS

Exxon LL5202.09 LLDPE, a barefoot 12 melt index, 0.924 density polyethylene was used as the carrier resin for preparation of all concentrate formulations. Exxon LL1001.09 LLDPE, a barefoot 1.0 melt index, 0.918 density film resin was selected as the matrix material for the preparation of all letdown compounds. Additive-free resins were used throughout this trial to minimize potential interactions with additive packages and to simplify the analysis task.

Samples of 275-0049 Diarylide Pigment Yellow 13 and 275-0570 Diarylide Pigment Yellow 83 were supplied by Sun Chemical.

Maleated resins included Eastman Chemical Epolene E-43 Modified Polyolefin and DuPont Fusabond MB-110D Modified Polyolefin. Epolene E-43 is a low molecular weight polypropylene wax graft-modified with relatively high levels of maleic anhydride (1-1.5 wt% MAH, Acid Number = 47-50). Number average and weight average molecular weights are 3900 and 9100 respectively. Fusabond MB-110D is a maleic anhydride graft-modified LLDPE resin with a melt index of 30 and an MAH functionality level higher than 0.5 wt%.

SAMPLE PREPARATION

The concentrate samples shown in Table 1 were prepared using a 1.0 liter laboratory HIDM and pelletized on a 1.25" Killion single screw extruder. All processing steps were carried out 150-165°C to minimize premature pigment decomposition.

Table 1. Diarylide PY13 and PY83 concentrate formulations

Concentrate sample	PY13 %	PY83 %	LLDPE %	Epolene E-43 %	Fusabond MD-110D, %
MB1	15.0	-	85.0	-	-
MB2	15.0	-	83.0	2.0	-
MB3	15.0	-	80.0	5.0	-
MB4	15.0	-	80.0	-	5.0
MB5	15.0	-	83.0	-	2.0
MB6	-	5.0	95.0	-	-
MB7	-	15.0	85.0	-	-

Subsequent letdowns were prepared by hand-mixing the pelletized concentrates with 1 MI LLDPE resin at the appropriate letdown level and melt compounding on the single screw extruder 150-165°C. Two letdowns (2.0 and 7.0 wt%) were prepared from each concentrate sample.

CONTROLLED THERMAL DECOMPOSITION

Thermal decomposition of the concentrate and compound samples was carried out using a Ray-Ran benchtop injection molding machine. This unit resembles an oversize melt index plastometer and is equipped with a heated barrel, die and plunger assembly. The capacity of the barrel is approximately 20 g.

Test materials were loaded into the heated barrel and compacted with the ram plunger. After 10 minutes dwell time, the charge was ejected onto a PET sheet and rapidly quench cooled in a room temperature platen press. Two 20 gram shots were prepared for each test formulation. After cooling, the plaque samples were cut by hand into 1x1 cm pieces and submitted for DCB analysis.

Concentrate samples MB1 through MB7 were treated at 240°C for 10 minutes to simulate excessive processing conditions during manufacture. These were designated MB1-240 through MB7-240. Dilute compounds 1A-7A and 1B-5B were treated at 240°C to simulate excessive processing conditions in typical end-use applications. These were designated 1A-240, 1A-270, etc. Selected samples were also subjected to the higher 270°C treatment (see Figure 2 for details).

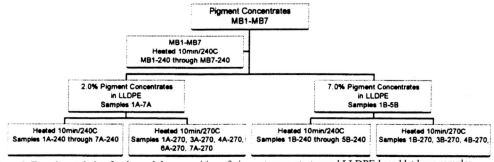

Figure 2. Experimental plan for thermal decomposition of pigment concentrates and LLDPE-based letdown samples.

3,3'-DCB ANALYSIS

Samples were analyzed for free 3,3'-DCB using a modification of the technique described by Az, Dewald and Schnaitmann.[2] Soxhlet toluene extractions were carried out under mild conditions to minimize pigment degradation and possible cleavage of imide/amide DCB adducts during the course of the extraction. The samples were analyzed for 3,3'-DCB with C18 reverse phase HPLC using UV detection. Analysis results are presented in Table 2. The calculated limit of detection for the method was approximately 0.2 ppm.

Table 2A. 3,3'-DCB analysis results

Sample Designation	Description	DCB, ppm
PY13	Diarylide Pigment PY13 (as received)	4.90
PY83	Diarylide Pigment PY83 (as received)	31.07
LLDPE	LLDPE Polyethylene Resin (as received)	none detected
MB1	15% PY13 in LLDPE	0.31
MB2	15% PY13 + 2% E-43 in LLDPE	0.54
MB3	15% PY13 + 5% E-43 in LLDPE	none detected
MB4	15% PY13 + 5% MD-1110D in LLDPE	0.21
MB5	15% PY13 + 2% MD-110D in LLDPE	0.88
MB6	5% PY83 in LLDPE	0.91
MB7	15% PY83 in LLDPE	1.41
1A	2% MB1 in LLDPE (0.3% PY13)	none detected
2A	2% MB2 in LLDPE (0.3% PY13)	none detected
3A	2% MB3 in LLDPE (0.3% PY13)	none detected
4A	2% MB4 in LLDPE (0.3% PY13)	none detected
5A	2% MB5 in LLDPE (0.3% PY13)	0.30
6A	2% MB6 in LLDPE (0.1% PY83)	none detected
7A	2% MB7 in LLDPE (0.3% PY83)	none detected
1B	7% MB1 in LLDPE (1.05% PY13)	none detected
2B	7% MB2 in LLDPE (1.05% PY13)	none detected
3B	7% MB3 in LLDPE (1.05% PY13)	0.27
4B	7% MB4 in LLDPE (1.05% PY13)	none detected
5B	7% MB5 in LLDPE (1.05% PY13)	0.44

RESULTS AND DISCUSSION

Preliminary FT-IR spectroscopy studies were made using maleated LLDPE doped with a slight excess of 3,3'-DCB. Samples were heated on a 150°C roll mill for 10 minutes. The extent of anhydride/amine reaction was inferred by the disappearance of the MAH carbonyl band[7] at 1789-1790 cm^{-1} (Figure 3). Discrete amide band increases were not observed, but broad changes in the 1740-1750 cm^{-1} region suggested a combination of amide, imide and carboxylic acid reaction products. Although the 3,3'-DCB diamine could lead to crosslinking reactions at low concentrations (i.e. by forming a bridge between two reactive maleic sites), no signs of gel formation were observed. It should be noted that while the infrared technique

Table 2B. 3,3'-DCB analysis results (samples kept at elevated temperatures)

10 min @ 240°C		10 min @ 270°C	
Sample	**DCB, ppm**	**Sample**	**DCB, ppm**
MB1-240	0.42		
MB2-240	0.62		
MB3-240	0.48		
MB4-240	0.97		
MB5-240	1.29		
MB6-240	3.06		
MB7-240	2.47		
1A-240	none detected	1A-270	5.20
2A-240	8.77		
3A-240	none detected	3A-270	9.76
4A-240	none detected	4A-270	7.89
5A-240	none detected		
6A-240	0.29	6A-270	1.04
7A-240	none detected	7A-270	3.22
1B-240	none detected	1B-270	5.91
2B-240	10.72		
3B-240	none detected	3B-270	5.48
4B-240	none detected	4B-270	5.12
5B-240	none detected		

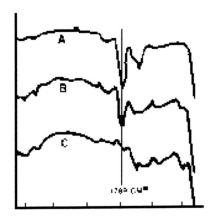

Figure 3. Infrared spectra of maleated LLDPE/3,3'-DCB blend as a function of reaction time at 150°C; A) T=0 min, B) T=5 min, C) T=10 min.

proved useful in examining trends in model resin systems, it was not well suited to the heavily pigmented compounds examined in subsequent trials.

HPLC analysis of Pigment Yellow 13 and Pigment Yellow 83 samples as received indicated that both contained residual quantities of free 3,3'-DCB, presumably left over from the manufacturing process. HPLC analysis gave a value of 4.9 ppm (average of two analyses) for Pigment Yellow 13 while Pigment Yellow 83 was assayed at 31.1 ppm. The manufacturer advises that typical 3,3'-DCB level in these pigment are approximately 2.5 and less than 50 ppm, respectively.

Interestingly, only one of the concentrate formulations made from these pigments (MB5) contained 3,3'-DCB levels higher than "expected" values calculated based on dilution of the raw pigment values. It is possible that a small quantity of 3,3'-DCB was released during the HIDM compounding, but this was difficult to confirm. It is also interesting to note that all of the masterbatch formulations examined contained less than 2 ppm free 3,3'-DCB.

Samples MB2 and MB5 contained higher DCB levels than the MB1 control, while samples MB3 and MB4 appeared to have marginally lower levels. This seems to suggests that low levels of maleated modifiers seem to have a detrimental effect on free DCB levels, possibly as a result of partial solvation of the pigment by the low molecular weight additive. If this were the case, however, one might expect to see a greater effect from E-43 than MD-110D and in fact the opposite is true.

When heated at 240°C for 10 minutes, all seven concentrate samples (MB1-240 through MB7-240) exhibited an increase in free 3,3'-DCB. Of the five PY13 formulations, the control (MB1-240) showed the lowest level, while MB5-240 contained the highest. In addition, the more thermally stable PY83 blends showed relatively large increases in DCB level. In this set of experiments, the MB6-240 (5% PY83) formulation yielded 3.06 ppm versus MB7-240 (15% PY83) which gave 2.47 ppm.

In general, the 2% letdown compounds exhibited only trace quantities of 3,3'-DCB (ND-0.29 ppm) under heat aging at 240°C. The only anomalous result was a high 8.75 ppm reading from sample 2A-240. This sample was not subjected to 270°C conditions, but other PY13 formulations showed significant increases in free 3,3'-DCB levels at the higher temperature. Both 3A-270 and 4A-270 exhibited higher levels than the 1A-270 control, again indicating that that the E-43 and MD-110D maleated components were contributing to 3,3'-DCB levels rather than reducing them. PY83 samples 6A-270 and 7A-270 showed lower overall increases in decomposition product, confirming that the pigment is more thermally stable than PY13 at typical use concentrations.

Of the 7% PY13 samples treated at 240°C (1B-240 through 5B-240), all but one exhibited free 3,3'-DCB levels below the limit of detection. The exception was sample 2B-240, which once again demonstrated exceptionally high levels (10.72 ppm). At 270°C, samples 1B, 3B and 4B yielded virtually identical results (5.91, 5.48 and 5.12 ppm, respectively), suggesting that here the MAH-modified E-43 and MD-110D additives provided little or no significant positive or negative contribution to 3,3'-DCB levels.

CONCLUSIONS

1. Both modified and unmodified PY13 concentrate formulations appear to be stable during routine compounding (extrusion temperatures up to 240°C and residence times of 10 minutes). DCB levels in the heat-aged PY13 concentrates all remain below 1.29 ppm. MAH-modified formulations contain more free 3,3'-DCB than the unmodified control.

2. Unmodified PY83 concentrates contain higher DCB levels than any of the PY13 samples. This is likely due to the high DCB content in the PY83 starting material. In addition, PY83 concentrates exhibited significant increases on heat aging with the 5% sample showing 3.06 ppm and the 15% sample showing 2.47 ppm.

3. All of the 2% and 7% PY13 and PY83 concentrate/LLDPE compounds exhibited DCB levels near or below the limit of detection (0.44 to <0.2 ppm) after initial low temperature compounding.

4. With one exception, all of the 7% PY13 and PY83 concentrate/LLDPE compounds showed negligible increases in 3,3'-DCB levels after treatment at 240°C. Levels increase significantly after treatment at 270°C. Pigment Yellow 83 appeared to be more thermally stable than Pigment 13 at typical loading levels encountered in polyolefin applications.

5. There is no indication that the MAH-modified additives studied here were effectively trapping free 3,3'-DCB as it was being created. On the contrary, it appears that the low molecular weight of the additives may be contributing to higher rates of 3,3'-DCB generation, possibly by contributing to solvation of the diarylide pigment itself. This tends to confirm previous evidence that the diarylide decomposition reaction can be extremely sensitive to concentration and the presence of low molecular weight fractions and additives in the polymer matrix.[2,4,8,9]

ACKNOWLEDGMENTS

We wish to acknowledge and thank Sun Chemical Pigment Division for their valuable assistance throughout. We would also like to thank Rushi Amin who conducted much of the preparatory work. Finally, we wish to acknowledge the support of the National Research Council of Canada Industrial Research Assistance Program for their financial support of this work.

REFERENCES

1. Letter to Customers from Hoechst Celanese Color & Surfactant Division, Coventry RI (April 18, 1990).
2. R. Az, B. Dewald, D. Schnaitmann, *Dyes and Pigments*, **15**, 1 (1991).
3. **IARC Monographs on the evaluation of carcinogenic risk of chemicals to man: 3,3'-dichlorobenzidine**, Vol. 4, 49 (1974).
4. F. Leuschner, *Toxicology Letters*, **2**, 253 (1978).
5. Ecological and Toxicological Association of the Dyestuffs Manufacturing Industry (ETAD) Information Notice No.2 Thermal Decomposition of Diarylide Pigments, September (1990).
6. Ecological and Toxicological Association of the Dyestuffs Manufacturing Industry (ETAD) Report T 2028-CA, On the Carcinogenic Potential of Diarylide Azo Pigments Based on 3,3'-Dichlorobenzidine, September (1990).
7. O. Laguna, J. P. Vigo, J. Taranco, J. L. Oteo, E. P. Collar, *Rev. Plast. Mod.*, **58**, 398 (1989).
8. Letter from the Dry Color Manufacturers Association Diarylide Pigments Committee (August 14, 1990).
9. Letter from PMS Corporate Technical Center to the United States Environmental Protection Agency (July 23, 1990).

Photoresponsive Polyurethane-Acrylate Copolymers

Eduardo A. Gonzalez de los Santos, M. J. Lozano-Gonzalez
Centro de Investigacion en Quimica Aplicada, Blvd. Enrique Reyna H. # 140, Saltillo Coah. c.p., 25100 Mexico
A. F. Johnson
Interdisciplinary Research Center in Polymer Science and Technology, School of Chemistry, University of Leeds, Leeds LS2 9JT, England

ABSTRACT

In an attempt to correlate the molecular structure of the spiropyrane merocyanine transformation with its photo-response (optical and mechanical properties), four spiropyrane and *bis*-spiropyrane molecules were examined. An optically transparent polyurethane-acrylate block copolymer was synthesized, containing the photo-chromic pigments. It is demonstrated that structural changes in photo-chromic pigments, can be utilized for modification and control of mechanical or optical properties of polymers and that these changes can be manipulated and predicted by computer molecular modelling methods.

INTRODUCTION

A large number of organic and inorganic materials which exhibit photochromism have been known for many years.[1] Such materials have been applied in many areas of technology to use them as dosimeter materials, light control filters, recording films in photography and decorative paints.

Now, application in 3D optical storage memory and lasers devices[2] or infrared sensitive spiropyranes,[3] are the focus of investigation in optical technology, which has raised interest in these photoresponsive materials.

Photochromism is the phenomenon where the absorption spectrum of a molecule or crystal changes reversibly when the sample is irradiated by light of a certain wavelength. When such a photoisomerization reaction is carried out in a polymer matrix instead of in solution, a strong decrease in the reaction rate (coloration = decoloration) is observed.[4] The polymeric effects are more strongly pronounced when the photochrome is chemically bonded to the polymer matrix. Such effect was attributed to the reduction of chain segment mobility.[5]

Additionally, in light-sensitive polymers containing photochromic compounds, proper irradiation leads to photoisomerization of the photochrome, and under certain conditions it also leads to conformational changes of the matrix which contain the photochrome. In solution, this effect is reflected in a change of viscosity properties after irradiation,[6] while in solid samples it is observed as a change in the macroscopic dimensions and mechanical characteristics of the sample.[7]

The present report describes the observation and measurement of the photo-optical and photo-mechanical responses on polyurethane-acrylate block copolymers in which a photochrome (spiropyrane or *bis*-spiropyrane) has been incorporated in the form of a simple solid solution or chemically bonded.

EXPERIMENTAL PROCEDURE

SYNTHESIS OF PHOTOCHROMIC COMPOUNDS

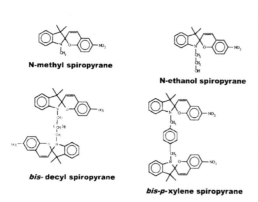

N-methyl spiropyrane

N-ethanol spiropyrane

bis- decyl spiropyrane

bis-p-xylene spiropyrane

The novel synthesis of spiropyranes and *bis*-spiropyrane compounds were described in a recent report,[8] which employs ultrasound as the energy source to promote the condensation reaction between the indoline or the *bis*-indoline and the desired salicylaldehyde, as well as the potential of the method which offers advantages over the conventional method. The photochromic compounds employed in this study are shown in Figure 1.

Figure 1. Photochromic compounds employed in the study.

SYNTHESIS OF POLYURETHANE-ACRYLATE BLOCK COPOLYMERS

For the synthesis of urethane-acrylate prepolymer, hexamethylene diisocyanate (HDI, 0.178 moles,) was placed in a reaction vessel, then 2-hydroxyethyl-acrylate (HEA, 0.178 moles) was added. The reaction temperature was kept at 45°C for 30 min., to avoid thermal polymerization through the vinyl groups, then a stoichiometric quantity of dry polypropylene glycol (PPG) (MW 725, hydroxyl number 147 mg of KOH/g (67.4 ml.)) was added along with dibutyl-tin-dilaurate as a catalyst

(0.674 ml., 1% wt). The mixture was heated at 70°C and stirred for 2 hrs. A white wax-like material was formed with a T_g at -47°C and a T_m of 40°C.

The thermally curable liquid mixtures were formulated from the above prepolymer by the required amount of methylmethacrylate (MMA, 32, 50 and 72 wt%) and azo-*bis*-isobutyronitrile (AIBN) catalyst. The mixture was cured for 15 min in the mold at 80°C. The photochromic materials were dissolved and added along with the MMA.

MEASUREMENT OF PHOTOCHROMIC RESPONSE

The maximum wavelength absorbance was measured for all photochromic pigments. Samples were irradiated with a specific UV wavelength of 325 nm for 30 s, which activated the photochromic compound in the polymer matrix. The material developed a characteristic color upon irradiation. Immediately after irradiation, the samples were scanned from 380 nm to 900 nm at 20°C. The visible spectrum was recorded using a UV/VIS Spectrophotometer Phillips Scientific PU8730.

The decoloration rates at the maximum absorption wavelength were also measured, using fresh specimens activated by the procedure described above. The specimens were monitored at the appropriate wavelength and the decoloration rate was measured for 100 min. at 20°C. The decoloration rate constants (k) were determinated assuming[9] that the fading is a first order reaction. Plots of ln k versus acrylic content for each photochromic compound were constructed.

MEASUREMENT OF PHOTO-MECHANICAL RESPONSE

A tensile tester (Instron Model 5564) was used to carry out measurements of the photomechanical response at constant length. Rectangular tensile samples (50 x 10 x 0.5 mm) were cut from the films, which were stored in the dark for 48 hrs, in order to allow any color to decay. The specimen was loaded with a constant stress load (80 g) allowing 10 minutes for relaxation. Then the sample was irradiated for 5 minutes with a UV lamp (325 nm) (Model P66100/9 Philip Harris). In order to remove any heat radiation from the lamp, a heat-absorbing filter was used (Model HG3 Melles Griot, Schott KG glass). After 5 minutes of irradiation, the lamp was switched off and the specimen was left in the dark for another 5 minutes. Stress changes with time were recorded for each light-dark cycle.

RESULTS AND DISCUSSION

DETERMINATION OF THE MAXIMUM WAVELENGTH.

Figure 2 shows the visible spectrum obtained after irradiation for each photochromic compound employed in the study at 37% of acrylate content. The maximum absorbance for each individual formulation after 30 sec. of irradiation is given in Figure 3. The general trend ob-

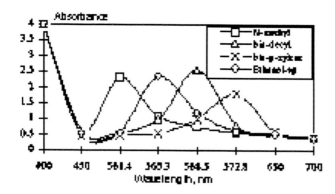

Figure 2. UV-Vis spectrum obtained after irradiation at 37% acrylate content.

served is that the lower the acrylate content the higher the absorbance. The lower the acrylate content the softer the material, consequently the photochrome has less steric restriction in the surroundings, making the photoisomerization easier. Also the *bis*-decyl spiropyrane series has slightly higher absorbance, within the 30 sec. of irradiation, compared with the rest of the series. No-

Table 1. Maximum wavelength absorbance for each photochromic compound

Compound	λ_{max}, nm
N-methyl spiropyrane	561.4
N-ethanol spiropyrane	565.3
bis-decyl spiropyrane	568.5
bis-p-xylene spiropyrane	572.8

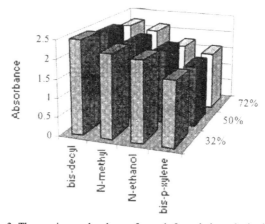

Figure 3. The maximum absorbance for each formulation, obtained at its maximum wavelength of absorbance.

tice that the *bis-p*-xylene spiropyrane series has the lower absorbance, even when this has twice the number of active centers than N-methyl spiropyrane or N-ethanol spiropyrane.

If it is *a priori* assumed that the larger the molecule the bigger the hindrance for the photochrome in the polymer matrix, then the dimensions of the molecules (Table 1) do not agree with the results observed in the Figure 3. Theoretical maximum lengths for the spiropyranes and the merocyanines were obtained by a computer-based modelling calculation[10] using SPARTAN V4.11. The *bis-p*- xylene spiropyrane has the higher difference between the two forms (3.83 Å), and the *bis*-decyl spiropyrane the lowest difference in length

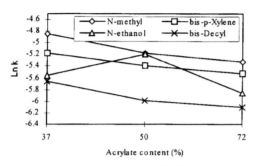

Figure 4. Decoloration rate constant (ln k) along the percentage composition for all photochromic compounds.

(0.77 Å). In this case, it is clear that the previous reasoning (the bigger the molecule the higher the hindrance in the polymer matrix) used for the first instance could be wrong, a better reasoning might be, the larger the transformation of the photochrome (close to open form) the higher the hindrance in the polymer matrix. The latter shows better agreement with the results observed.

DETERMINATION OF DECOLORATION RATE

The decoloration rate constants (ln k), estimated as a first order reaction, against the percent of acrylate content are shown in Figure 4. In general, the lower the acrylate content the faster the decoloration rate. The observed fading trend is as follow: N-methyl > *bis*-p-xylene > N-ethanol > *bis*-decyl spiropyrane. The N-methyl merocyanine has the faster decoloration rate as expected, the N-ethanol merocyanine present a slowness in almost 0.7 units (at 37%) as it is chemically bonded as a side group. Out of trend is the noticeable high value of fading rate at 50% acrylate content. During phase inversion (around 50/50) and when the two components are present as a co-continuous phase, it is possible that important interactions, observed for the photochrome chemically linked to the polyurethane backbone and the acrylate phase, disappear. This leaves the photochrome free to achieve the transformation (open to closed form) without restrictions. At 76% the fading rate follows the trend reasonably well.

For the *bis*-spiropyranes case, the *bis*-p-xylene merocyanine has the faster fading rate compared with the *bis*-decyl merocyanine. *A priori* one would expect that the *bis*-decyl merocyanine would have a faster fading rate, taking into account that this molecule is bigger (24.1 Å) than the *bis*-p-xylene merocyanine (20.4 Å), therefore it will experience more stress in the polymer matrix, so the recovery rate back to the spiropyrane form would be faster. An additional effect is obviously involved.

From the modelling calculations, it was observed that the *bis*-decyl merocyanine has a higher dipole moment (45.4 Debyes) than *bis*-p-xylene merocyanine (27.9 Debyes). Since the merocyanine form develops a large dipolar moment, several times higher than the spiropyrane, it can be anticipate that this dipole interacts with polar polymer matrix. Hence, the bigger the dipole the higher the interaction. If this occur, the transformation from the merocyanine to the spiropyrane will experience an significant delay,[10,11] consequently this delay will be noticed in the fading rate.

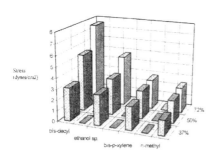

Figure 5. Graphical representation for the photomechanical response.

Table 2. Maximum values of volume obtained by computer-based modelling for spiropyranes and merocyanines (in $\mathring{A}^3$)

Compound	Spiropyrane	Merocyanine	Difference
N-methyl	431.43	280.56	150.87
N-ethanol	440.67	275.69	164.98
bis-p-xylene	1087.30	734.66	352.64
bis-decyl	1462.13	1199.89	262.24

DETERMINATION OF PHOTO-MECHANICAL RESPONSE

A graphical representation for the measurements of the photo-mechanical response is shown in Figure 5. Upon irradiation, the initial stress applied to the sample decreases indicating an expansion of the sample, while in the dark, length recovery takes place by a subsequent contraction indicated by an increase in stress, the process can be repeated many times. Blair and Pogue,[7] reported a photoexpansion in samples with photochromic material.

It can be seen also that the higher the acrylate content the larger the photo-mechanical response. The acrylate is a rigid homopolymer, so can transmit the work more easy than the polyurethane (soft material), which can readily absorb the work applied by the photochromic transformation.

As a control, polyurethane-acrylate copolymers without photo-chromic agents were subjected to the same photo-mechanical test. No significant photomechanical response was observed.

From the volume calculations (Table 2) obtained by computer-modelling,[7] it was assumed that the *bis*-p-xylene would induce more spatial disruption in the polymer matrix, as the transformation from spiropyrane to merocyanine reflects more spatial changes (352 $\mathring{A}^3$), it would be expected to induce the greatest photo-mechanical effect. However, Figure 5 shows that the greatest photo-mechanical response (expansion) is observed for *bis*-decyl spiropyrane. The spatial disruption is not the main driving force for the photo-mechanical effect in this case.

There are two postulated mechanisms for photomechanical response, which taken together can explain the photochromic effect seen in this work. The first by Lovrien,[14] suggest that if a polymer interacts with some photochromic chromophore, it may undergo a conformational change when irradiated because the interactions between the polymer and the chromophore suffer changes. The second one, postulated by Matejka[11] and Irie,[12] suggest that some photochromic molecules have a big dipolar moment, hence they tend to orientate in parallel; compact coil conformations are preferred. Hence, some reduction in the entropy of the system will produce an expansion in the sample.

From the theoretical energy calculation, the *bis*-decyl showed the biggest dipolar moment (45.4 Debyes) compared with *bis*-p-xylene (27.9 Debyes), and the charge-charge interaction showed the same trend, -88.1 Kcal and -79.8 Kcal respectively. Taking this into account it is possible that *bis*-decyl, having less spatial disruption (262.24 Å^3) compared with *bis*-p-xylene (352.64 Å^3), has yields the greatest photo-mechanical response, according to the two postulated mechanism mentioned above.

The ethanol spiropyrane showed a higher response than the *bis*-p-xylene and *n*-methyl spiropyrane. However, the response is no greater than that shown by *bis*-decyl spiropyrane, suggesting that there is a compromise between the dipolar formation, spatial disruptance and the effect obtained by attaching the photochromic material to the polymer backbone.

CONCLUSIONS

In general, it was observed that the higher the acrylate content of the copolymer, the more photo-response is observed. Rigid homopolymers can easier transmit the work than soft materials. It was also demonstrated that the spatial disruption generated by the photo-transformation of the photochrome is not the main driving force for the photo-mechanical and photo-optical effects. It is important to take into account that stereo-electronic interactions between the polymer matrix and the photochrome exist, mainly for the merocyanine form.

From this study, it was concluded that an ideal photochromic spiropyrane molecule for inducing a large photo-mechanical effect, would be one that exhibits high spatial disruption, form strong dipoles in the merocyanine form, and is preferably bonded to the polymer backbone.

The response prediction can be done effortlessly by computer modelling in order to establish the properties of the molecule in both forms (spiro and merocyanine). It also can help to understand, predict and even anticipate the photo-mechanical behavior of such compounds in a determinated polymer matrix.

ACKNOWLEDGMENTS

Our gratitude to the IRC in Polymer Science and Technology for support of this project. The author E. A. Gonzalez also give thanks to the National Research Council of Mexico (CONACYT).

REFERENCES

1 R. C. Bertelson, **Photochromism**, G. H. Brown, Ed., *Wiley*, Chap III (1971), New York.
2 Ron Dagani, *Chemical & Engineering News*, September 23, (1996). Dimitri A. Parthenopoulos, Peter M. Renetzepis, *Science*, **245**, 843 (1989).
3 Seto Juníetsu, *Infrared Absorbing Dyes*, M. Matsuoka, Ed., *Plenum Press*, Chap. 7 (1990), New York.
4 G. Smets, J. Breken, M. Irie, *Pure Appl. Chem.*, **50**, 845 (1978).
5 G. Smets, *Pure Appl. Chem.*, **42**, 509 (1975)., *Pure Appl. Chem., Macromol. Chem.*, **8**, 357 (1973).
6 M. Ire, A. Menju, K. Hayashi, G. Smets, *Polym. Lett.*, **17**, 29, 1979. M. Ire,, Menju, A., Hayashi, K., *Macromolecules,* 12, 1178, 1979.
7 F. Angolini, F. P. Gay, *Macromolecules*, **3**, 349, 1970. G. Smets, G. Evens, *Pure Appl. Chem. Suppl. Macromol. Chem.,* **8**, 357, 1973. G. Smets, J. Braeker, M. Irie, *Pure Appl. Chem.*, **50**, 845, 1978. H. S. Blair, H. I. Pogue, *Polymer,* **20**, 99, 1979. L. Matejka, K. Dusek, M. Ilavsky, *Polym. Bull.*, **1**, 659, 1979. C.D. Eisenbach, Prep IUPAC 26th International Symposium on Macromolecules, Mainz Vol. 1, 293, 1979.
8 E. A. Gonzalez, S. Torres, L. Vazquez, *Synthetic Comm.*, **25**, 105, 1995. E. A. Gonzalez, M. J. Lozano, *Synthetic Comm., in press*. E. A. Gonzalez, M. J. Lozano, S. Torres, **Mex. Pat. 947449**, Sept., 24th, 1994.
9 R. Guglielmetti, *Photochromism Molecules and Structures*, Ed. H. Durr, H. Bouas-Laurent, *Elsevier Science Pub.,* 1990, New York.
10 Spartan V4.11, *Wavefunction, In*c. (1996), Performed On Digital, Alpha station 255,300 Mhz.
11 L. Matejka, K. Dusek,, *Makromol. Chem.*, **182**, 3223 (1981).
12 M. Irie, T. Suzuki, *Makromol. Chem. Rapid Commun.*, **8**, 607 (1987)
13 H. S. Blair, H. I. Pogue, *Polymer*, **23**, 779, 1982.
14 R. Lovrien, II Reversible Photochem. Processes Symp., Dayton, 26, AD 653030, 1967.

Safety, Health and Environmental Regulatory Affairs for Colorants used in the Plastics Industry

Hugh M. Smith
Sun Chemical Corporation

ABSTRACT

The subject of safety, health and environmental affairs is today so volatile and capable of change, that it is anticipated that some of the following information will be obsolete in the months following this publication. This paper then presents only a "snapshot" of the state of affairs existing at the time of writing, and not perceived as a static treatise on a most dynamic topic.

INTRODUCTION

In order to better understand the major issues facing today's dye and pigment industries, five general principles are presented for consideration:

THE NEED FOR WORKPLACE AND ENVIRONMENTAL REGULATIONS... THE LEGACY OF OUR FATHERS

As one who grew up in an industrial society where the safety and health of the individual, and the related impairment of the environment were occasionally submerged beneath the paramount goals of efficiency and profitability, the birth of today's regulatory maze was both logical and overdue. The "Legacy of Our Fathers" must thus be acknowledged as typical of at least a small portion of past chemical industry, and with it, the dye and pigment industries.

Two unfortunate, but well documented examples that come to mind are 1) the multiple cases of bladder cancer which occurred some thirty to forty years ago in an Ohio dyestuff intermediates facility, manufacturing the human carcinogen, benzidine, and 2) a strikingly similar incident in Georgia, during the same time period, involving the manufacture of another human carcinogen, beta-naphthylamine, for dyestuff production. Ironically, at the very time during which employees were being inadvertently exposed to the two carcinogens, the danger of these agents to human health was well known to medical specialists, but apparently was not known to the manufacturers concerned. National standards for worker protection were obviously overdue. Today we are thankful for a major paradigm shift which has occurred in recognition of the imperative of safety, health and environmental protection, in most manufacturing industries within the developed world.

THE MULTIPLICITY OF REGULATIONS, AND THE PROBLEMS OF INFORMATION OVERLOAD

As a result of this paradigm shift, and ever-increasing number of worker, consumer and environmental protection measures have been implemented in the United States by means of national, state, and occasionally, even by city ordinances. Sometimes, following a change in national administration, when political appointees have been known to gravitate from federal agencies to states sympathetic to their previous regulatory agenda, state regulations may be introduced well in advance of a federal counterpart. Again, from time to time it is not unusual to see individual state-to-state variations in regulatory language covering a common issue, to the point where detailed understanding of the difference frequently requires specialized insight beyond the in-house capabilities of most smaller dye and pigment manufacturing companies. And if we add to this often confused situation, a consideration of the export activities of manufacturers, processors and users of dyes and pigments, the many disparate requirements overseas are such as to provide a veritable maze; through which today's manufacturer, exporter or importer or consumer of dye and/or pigment has to pass. Currently, the sheer multiplicity of safety, health and environmental regulations encountered, represent an information overload of truly enormous proportions, if there will be an resolution to this dilemma, it surely must come from a second paradigm shift, involving regulators and regulated communities alike, as well as a concerned public, in finally accepting the necessity for regulatory harmonization. This would create a level "playing field," by which worker, consumer and environment can equally and adequately be protected.

GLOBALIZATION OF LOCAL ISSUES, AND THE PROBLEM OF THE SHRINKING UNIVERSE

At first sight, the probability of local workplace and environmental constraints in one part of the country impacting dye and pigment business thousands of mile away in another may seem remote. In recent years, however, the borders of regulatory enforcement have become somewhat blurred, due to business rather than regulatory requirements. Two examples from the pigments industry will serve to illustrate the point. Model legislation, developed by the Source Reduction Council of the Coalition of North Eastern State Governors (CONEG), was enacted in the early nineteen nineties by several states (through not yet by the federal government), setting limits on trace amounts of the four heavy metals — mercury, cadmium, lead, and hexavalent chromium, in packaging material. As a result, pigment manufacturers in the US were required to comply with this limitation in products sold into the packaging ink, and ultimately packaging industries. But more recently, some European manufacturers selling pigments to European packaging ink houses who in turn, sell products to European packaging concerns who may export their packages into the United States, are requiring the same restriction on heavy metal impurity levels which was originally conceived of as a purely local, US state issue.

Again, commercialization of a new dye or pigment substance in the United States, requires premanufacturing notification to USEPA, before the new substance can be entered on the Toxic Substances "Inventory," and legally manufactured/marketed/imported. But in some recent instances, it is not unusual for a dye or pigment maker to be required to furnish to the customer proof of presence on the Inventories of Japan, Australia, Korea or the Philippines, or wherever the user plans to export his downstream products.

SUBJECTIVITY IN RISK PERCEPTION, AND THE PROBLEM OF BALANCE

Today, we live in a media-molded society, where newsworthiness is often judged more significant than the cold prose of scientific assessment. We have already discussed the paradigm shift in industrial values to one where protection of the environment and human health is now perceived as more important than the profitability and efficiency of a business. But we must also take into account a radical view espoused by community and national activist bodies, and beloved by the media, that industry must never be trusted to self-police its boundaries, that increased regulatory enforcement is the only way to hold industry accountable for its practices, and that use of the media is legitimate to popularize the activist viewpoint, even if somewhat exaggerated claims are made against industry. In such vein, legally permitted discharge of dye or pigment industrial effluent for treatment by publicly operated treatment works (POTW) is invariably termed "dumping" of "pollution" and even the handling of "tainted toxins" by the media.

As might be expected, such "checks and balances" upon industrial practice are viewed as either an unbridled abuse of power (by industry), or a necessary evil (by environmentalists). Where does truth lie? Certainly, not as the media usually portray it! Rather, the truth is often hidden in scientific terminology that is difficult for the public to understand. It is hoped however, that the fledgling Science of Risk Assessment and Risk Communication will sometime advance to the point where objective truth in safety, health and environmental issues impacting dyes and pigments no longer need an interpreter.

ECONOMIC ISSUES, THE PROBLEMS OF DEEP POCKET PERCEPTION AND THE UNLEVEL PLAYING FIELD

Dyes and pigments in the United States are regulated today from their birth, through their processing, to their use, and ultimate fate within the environment. Non-compliance with these regulations is no longer considered a trivial offense, which companies can easily afford to pay, as a normal cost of doing business. Several dye and pigment manufacturers have already found this out, much to their dismay! Several million dollars in fines are unfortunately now commonplace! It is clear that the US chemical industry, is perceived by the regulatory agencies, as having "deep pockets" containing limitless sums of money to pay for enforcement violations. Sadly, however, such is not the case with the majority of US dye and pigment manufacturers, for fines of this magnitude could easily cripple their business. A second economic fact of life has to do with importation of dyes and pigments made offshore without the same restrictions levied on US manufacturers. In other words, undue economic burden from US regulatory requirements upon US dye and pigment manufacturers could eventually make offshore products, providing their quality and impurity profiles are comparable, more and more attractive to the US customers.

PREMANUFACTURING CONSIDERATIONS

In 1976, the US Congress enacted the Toxic Substances Control Act, popularly known as TSCA (PL94-469, 15 USC 2601 et seq.). Motivation for this action stemmed in part from awareness of widespread persistence of polychlorinated biphenyls (PCBs) and chlorofluoro-carbons (CFCs), in the environment, and the realization that EPA must be empowered to control risks associated with hazardous and toxic chemicals.

Under Section 5 of TSCA, EPA is empowered to assess the safety of all new substances, including dyes, pigments, raw materials or intermediates or additives, before their manufacture and/or importation into the United States. By definition, a new substance is one that does not appear on the TSCA *Inventory of Existing Chemical Substances*. In order to determine whether or not a Pre-manufacturing Notification (PMN) request must be filed, the non-confidential portion of the TSCA Inventory should first be consulted, and if necessary, a bona fide letter of intent to manufacture the substance submitted, as a trigger to EPA to search the Confidential portion of the TSCA Inventory for the substance in question. After determining that the substance in question is truly "new," and does not qualify for exemption from PMN review process (e.g., low volume, R&D only, impurity, by-product, site-limited intermediate, export only, etc.), a $2500 fee is submitted to the Agency, together with a comprehensive profile on the substance, including projected manufacturing plans, and toxicological reports, if available.

Following a 90-day review period, and typically several phone discussions between submitter and agency reviewers, EPA will either decide to regulate manufacture of the substance, by means of a "Section 5(e) order," or will raise no objection to manufacture. Within 30 days of first manufacture or importation, however, the submitting company must file an NOC (Notice of Commencement to Manufacture) with the Agency, after which EPA will add the new substance to the TSCA Inventory. And once the new substance appears on the TSCA Inventory, other companies are at liberty to manufacture or import it into the United States, provided that the Agency does not impose additional restrictions through imposition of a SNUR (Significant New Use Rule).

Today, a number of new dye and pigment products have been placed on the TSCA Inventory after PMN review. In addition, a significant number of pigment "additives," used in association with pigments to enhance their working properties (e.g., the aluminum salt of quinacridone sulfonic acid, or phthalimidomethylated copper phthalocyanine, both used for rheology enhancement in paint and ink systems respectively), have been synthesized and entered on the TSCA Inventory. Naturally, this area is one rife with confidential business information, but a careful reading of the recent patent literature should prove helpful in understanding the type and chemistry that may be involved.

PRODUCT SAFETY ISSUES

To manufacture and sell dyes and pigments today, several critical considerations must be reviewed by the manufacturer, importer and purchaser: the inherent toxicity of the dye or pigment, any toxic impurities which may be present, or could be produced upon breakdown of the product and the possibility of product misuse leading to undesirable effects produced on human health or the environment from over exposure. This area of concern is rightly called "product safety," and is a new discipline that has emerged in the United States following promulgation of TSCA. In general, the inherent safety of most classes of dye and pigment is well attested.[1,5] but several of the exceptions are noted below:

INADVERTENT IMPURITIES

Included in this area are the inadvertent presence of trace impurities in the product, e.g., PCB's in diarylide and phthalocyanine pigments, polychlorinated dibenzodioxins or dibenzofurans (PCDDs and PCDFs), in chloranil-derived dyes and pigments, and "CONEG Heavy Metals" in packaging ink grade dyes and pigments. Careful attention to the manufacturing process, including raw materials, solvents, and materials of construction used in the manufacturing equipment, together with exhaustive analyses has reduced such issues to controllable levels, as illustrated below:

Benzidine.   In past years where pigments, synthesized from benzidine were found to contain trace but detectable levels of this carcinogen, the products were discontinued from commerce and are no longer made.

Cyanide.   In former years, salt milling of phthalocyanine "crudes," using salt containing sodium ferrocyanide as an anti-caking agent, produced pigments containing analytically detectable amounts of cyanide. This practice has been discontinued in the US.

PCBs and PCDDs.   In Diarylide Yellow manufacture, the use of formate buffer to enhance pigment transparency for offset ink was discontinued in the nineteen seventies, when it was realized that polychlorinated biphenyls were being produced as an undesirable by-product of the coupling reaction. In Phthalocyanine Blue "crude" synthesis, the commonly used solvent, trichlorobenzene, was also discontinued in the US, as a potential source of poly-chlorinated biphenyls. More recently, use of chloranil, manufactured from chlorinated phenols, has been discontinued in the synthesis of dioxazine violet "crude" and sulfonated dioxazine acid dyes, so as to minimize by-product formation of polychlorinated dibenzodioxins and dibenzofurans (PCDDs/PCDFs). A new grade of high purity chloranil is now produced from hydroquinone for dye and pigment manufacture.

PRODUCT BREAKDOWN

The possibility of product breakdown to regenerate undesirable starting materials can be illustrated in examples from both dyes and pigments:

Benzidine Dyes.   The well-documented metabolic breakdown of benzidine dyes to the starting material and human carcinogen, benzidine, is today recognized by the dyestuff industries in both Europe and the United States.[1] Because of this, a voluntary withdrawal of this type of dyestuff was enacted several years ago in the US, and more recently, in Europe. Today, however, it is still possible to import benzidine dyes into the US from Mexico and India, and regulatory action appears to be necessary even by such a simple stratagem as deleting the dyes in question from the TSCA Inventory.

Diarylide Pigments.   A second example of product breakdown is that of diarylide pigments, derived from the animal carcinogen, 3,3'-dichlorobenzidine (DCB), by heating above 200°C in certain polymers and waxes, that partially solubilize the pigment and facilitate breakdown to DCB.[2] Because of the breakdown potential, this class of pigment is no longer recommended for the processing of plastics, such as polypropylene, polyamide, and polyesters, at temperatures in excess of 200°C.

PRODUCT TOXICITY

A few dyes and pigments are considered to be carcinogenic, by US regulatory agencies. Benzidine dyes have already been mentioned. Lead chromate pigments contain both lead and hexavalent chromium, and as such are defined by EPA as carcinogenic. Experimentally, however, lead chromate pigments have been found to be non-mutagenic and non-carcinogenic, due, no doubt, to their extremely low solubility.[3]

Certain dyes and pigments are recognized as skin and eye irritants. For example, diarylide pigments are sometimes treated with primary aliphatic amines to enhance dispersibility in publication gravure ink systems; and since the amine treatment agents are themselves known skin and eye irritants, it is not surprising that some of today's commercial pigments may require careful handling as slight irritants. A related area, currently under consideration by FDA, is that of skin sensitization from wood rosin, used in some cosmetic colorants.

SELF-HEATING OF DRY POWDERS

A separate issue that is receiving much attention at the present time, is the phenomenon known as self-heating. Certain dry pigments, including some heavily resinated diarylide yellows, metallized monoazos and black iron oxides, may exhibit an internal heating phenomenon to temperatures in excess of 200°C, when maintained over a 24 hour period at a temperature of 140°C. Such products are designated as self-heating substances, n.o.s., UN3088 and 3190, Packing Groups II and III, according to DOT requirements, which in turn are based upon United Nations Recommendations on the Transportation of Dangerous Goods.[4] Such designation mandates special labeling, and reference on the Material Safety Data Sheet (MSDS). Presently, there is no universal method for predicting self-heating properties. This would appear to be an excellent field for further research.

DEFLAGRATION

A related, but different area of concern is that of deflagration, which is the ability of a dye or pigment to support its own combustion under fire conditions without the necessity of an external source of oxygen. An excellent example of a deflagrating pigment is Dinitraniline Orange (C.I. Pigment Orange 5), a colorant which includes two nitro groups in its molecule. Fires involving such pigments must be handled carefully, even after dousing with water, as they have been sometimes perceived as apparently extinguished, only to progressively self heat, through a dry stage, to a smoldering stage, to reburn condition.

REACTIVITY CONSIDERATIONS

Most dyes and pigments are unreactive. One exception, however, is lead chromate, which can act as an oxidizer in the presence of certain monoazo pigments, and following intimate mixing, produce fires.[3]

MISCELLANEOUS

Two concerns which must be guarded against include accumulation of a static charge of electricity during transfer or processing operations on powdered dyes and pigments.[3,5] Such a charge can give rise to spark conditions, and thence fires. The possibility of dust explosions must also be considered. For example, before a new dye or pigment product is spray dried, it is good practice to carry out a dust cloud/dust layer assessment, confirm suitability for this operation.

PRODUCT SAFETY ISSUES

Four major workplace regulatory events have impacted the US dye and pigment industries:

OSHA'S WORKPLACE STANDARDS

Firstly (in 1970), the introduction of federal OSHA workplace standards (29 CFR 1910), codifying and requiring a multitude of working practices, by which dyes and pigments could be safely made.

OSHA'S CARCINOGEN STANDARDS

Secondly (in the mid-nineteen seventies), specific OSHA standards, for the safe handling of thirteen human and animal carcinogens. Included in the list is dichloro-benzidine, a key raw material for the manufacture of diarylide pigments, regulated under 29 CFR 1910.1007.

THE HAZARD COMMUNICATION STANDARD

Thirdly, in the nineteen eighties, as a result of the "Right to Know" movement for greater public awareness of the toxic and hazardous properties of materials to which workers might be exposed, OSHA's Hazard Communication Standard was promulgated as 29 CFR 1910.1200.

THE PROCESS SAFETY STANDARD

Lastly, in the nineteen nineties, OSHA's Process Safety Standard was introduced as 29 CFR 1910.119, requiring hazard analysis and process safety assessment of selected hazardous substances. For example, use of methanol in dye and pigment syntheses now requires an extensive process safety analysis and audit.

As a result of such workplace regulatory developments, hazard assessment of all materials used in U.S. dye and pigment workplaces is now commonplace, understanding of safe handling practices, including proper engineering controls and personal protective procedures, is widespread, and literacy in reading the freely available Material Safety Data Sheets is usually high. Compared with typical practices in the nineteen sixties and before, it is obvious that workplace safety today in the U.S., pertaining to the manufacture and use of dyes and pigments has undergone significant improvement.

The reader is encouraged to read the recent CPMA booklet, entitled, Safe Handling of Pigments,[3] for further treatment of this topic. A second edition is now under development, as well as multi-international versions, each customized with regard to prevailing national regulations.

THE ENVIRONMENT

AIR EMISSIONS

In 1990, Congress passed major amendments to the Clean Air Act (PL 101-549; 42 USC 7401), expressing serious dissatisfaction with EPA's prior regulation of airborne toxins, and imposing upon the Agency a requirement to regulate 189 listed hazardous air pollutants. Included in the list is dichlorobenzidine (DCB), a key raw material for Diarylide Yellow pigments. Today, the sole U.S. DCB manufacturer is required to control air emissions of DCB using Maximum Achievable Control Technology (MACT). At present, regulation of air emissions from dyes and pigments manufacturing plants is typically governed by state requirements, where raw materials as well as particulate matter from dyes or pigments production, are regulated. Because of this, it is customary for state approval to be obtained, before any new dye or pigment type can be introduced to a given manufacturing facility. It is also anticipated, that in the near future, MACT requirements will be placed on the control of volatile solvents used in the dye and pigment manufacturing industries.

WATER DISCHARGE ISSUES

Turning now to the issue of waste water release from dye and pigment manufacturing and using facilities, four major issues should be considered. Before doing so, it should be understood that in the U.S., waste water may be discharged directly into navigable waters, following pretreatment to levels proscribed by an NPDES (National Pollution Discharge Elimination System) permit, set by EPA or their state designee. Many dye, and some pigment manufacturing facilities have NPDES permits. Alternatively, many U.S. pigment facilities discharge their industrial wastewater into POTWs (Publicly Operated Treatment Works), which regulate their influent to suitable "pretreatment" levels, thus enabling the POTW to meet their NPDES permit.

One long-standing issue is the concern on the part of some POTWs, particularly on the west coast, to limit the quantity of copper phthalocyanine in wash water from clean up of flexographic ink presses. Although the copper found in the wash water is in an insoluble, tightly molecularly bound form, the use of EPA's documented analytical technique for copper determination uses strong nitric acid, which destroys the pigment molecule, unbinds the metal in the process, and measures it as if it were the soluble copper ion.

In addition to conventional effluent discharge limits for parameters such as Biological Oxygen Demand (BOD), Total Suspended Solids (TSS), Acidity (pH), Fecal Coliform, or Oil and Grease, a second, regulatory restriction for organic pigment and dye manufacturing facilities is the Organic Chemicals, Plastics and Synthetic Fibers Pretreatment Regulation (OCPSF, 40 CFR 414.80-85). This specifies pretreatment levels for a long list of organic and inorganic substances, regardless of direct discharge or discharge to a POTW. There have been instances in which compliance has required companies to address some unusual problems. One instance which comes to mind, is a recent problem, encountered by some U.S. azo dye and pigment makers involving trace but detectable levels of toluene in their waste water above prescribed OCPSF limits. Since toluene was not known to be present in the facilities, an exhaustive search unearthed the reason for the overage. It was found that a major manufacturer of arylide couplers had been using toluene as the reaction solvent during their synthesis, and residual solvent was adhering to the coupler, thus finding its way into the wastewater.

A third wastewater discharge issue is concerned with the ultimate fate of the sludge generated by a POTW. In cases where the POTW land-applies the sludge for agricultural purposes, new restrictions on specific heavy metals now make it necessary for some POTWs to place further restrictions upon industrial users. Specifically, this issue is now restricting the manufacture of molybdated pigments, and could possibly impact phthalocyanine dyes and pigments made with molybdenum oxide as catalyst.

A fourth wastewater discharge issue has recently been proposed as a new regulation by USEPA.[7] Responding to a legal challenge by the Environmental Defense Fund, EPA has analyzed and evaluated the characteristics of wastewater discharge from a cross section of dye and pigment plants, and has determined that wastewater and wastewater treatment sludge, from the production of azo dyes and azo pigments will be regulated as hazardous wastes under Sub-title C of the Resource Conservation Recover Act (RCRA). Both the dyes and pigments industries are strenuously opposed to the proposed rule, and have pointed out that while the incremental risk (in terms of cancer cases avoided) is near zero, two-thirds of U.S. pigments and dyes manufacturing facilities may incur significant costs, and one-quarter may face closure as a result.[8]

SOLID AND HAZARDOUS WASTE ISSUES

Under 40 CFR 261, SubPart C, a hazardous waste is one that exhibits ignitability, corrosivity, reactivity or toxicity. The last mentioned characteristic is evaluated by the Toxicity Characteristic Leaching Procedure (TCLP). The procedure checks the presence of 39 organic and several inorganic substances, one of which is barium. This means that manufactured barium salt pigments may contain regulated levels of acid leachable barium, which may render any resulting waste hazardous.

An important issue, is the need for educating many U.S. waste haulers, into recognizing that merely because a dye or pigment waste is highly colored, does not render it hazardous. This truth, unfortunately is still not well known enough, even today.

A further issue, impacting some disazo dyes and diarylide pigments, is that of environmental fate. Is it conceivable that under worst case conditions, waste dye or pigment might break down to the component starting amine? EPA is currently considering this issue, and contrary evidence regarding the non-biodegradability of diarylide pigments has recently been provided to the Agency. Environmental fate has also been a consideration for recyclers or printed material, colored plastics and other products containing dye and pigment.

THE TOXIC RELEASE INVENTORY

Each year, U.S. dye and pigment manufacturers are required to submit two itemized lists to local, state and federal authorities. Firstly, specific information on the quantity and location of OSHA-hazardous substances located at each facility; and secondly, a record of listed toxic chemical releases to air, water and waste. The latter listing is known as the Toxic Release Inventory (TRI), and is deliberately given public access by the environmental agencies. As might be expected, debate on the significance of the TRI data sometimes follows "party" lines, with manufacturers perhaps claiming that the releases are modest for the type of processes involved, are all within legally permitted limits, and are often declining each year, due to pollution prevention activities within their facilities. On the other hand, many environmental groups perceive quite another story in the same numbers, and each year translate "legally permitted" as "dumped" or "polluted," believing that the only good level of release (pollution) is zero. Such a dialogue would seem healthy for U.S. industry, were it not for the efforts of the media, who 1) often take an anti-industry bias, 2) may mislead the public, and 3) are ultimately responsible for a considerable financial outlay by industry, in an attempt to set this and similar records straight.

Within the U.S. dye industry, such a responsive role is played by the U.S. Operating Committee of the Ecological and Toxicological Association of Dyestuff Manufacturers (ETAD), while the North American Pigments Industry is ably served by the Color Pigments Manufacturers' Association (CPMA). One notable outcome in this area has been the recent "delisting" of copper phthalocyanine pigments from inclusion in the TRI Inventory. Among inorganic pigments which must be reported for the TRI, are pigments containing lead, cadmium, chromium, cobalt, nickel, and zinc.

ANALYTICAL COMPLIANCE

Over the years, the U.S. dye and pigment industries have often perceived themselves as targeted by Federal and State Agencies. And sometimes, "technology forcing" has been applied, by requiring industry to minimize environmental emissions or workplace levels of certain substances down to specified limits, without the regulating body having a clear idea as to the specific Control Technology necessary to achieve such limits; or, equally importantly, how to reliably analyze for the substances in question, in an industrial matrix, down to the levels required by regulation. As a consequence, recent analytical technology in both dye and pigment industries, has been mainly focused on methods development, in such areas as quantitating PCB impurities in diarylide and phthalocyanine pigments, CONEG heavy metal impurities in packaging grade dyes and pigments, and OCPSF listed substances in dye and pigment wastewater, etc.

CONCLUSIONS

As will now be apparent to the reader, a careful review of health, safety and environmental affairs pertaining to the U.S. dyes and pigments industries reveals major change and growing complexity over the last twenty-five years. In fact, regulatory concern can possibly be considered as the major force which has shaped today's industrial workplace and environment. With continued focus on dyes and pigments assured over the foreseeable future by state and federal agencies, it is certain that this area of endeavor will continue to merit close attention.

REFERENCES

1.   BNA, *Chemical Regulation Reporter*, 8-9, April 1, 1994.
2.   R. Az, B. Dewald, D. Schnaitmann, *Dyes and Pigments*, **15**, 1991.
3.   **Color Pigments Manufacturing Association, Safe Handling of Pigments**, First Edition, CPMA, Alexandria, VA, 1993.
4.   **United Nations, Recommendations on the Transport of Dangerous Goods**, *United Nations*, New York, 1990.
5.   National Fire Protection Association (USA), **Prevention of Fire and Dust Explosions in the Chemical, Dye, Pharmaceutical and Plastics Industries**, *NFPA*, 1988.
6.   H.M. Smith, The Toxicology of Organic Pigments, *American Paint and Coatings Journal*, October, 1993.
7.   *Hazardous Waste Management System; Identification and Listing of Hazardous Waste; Dye and Pigments Industries*, Federal Register, 59, December 22, 1994, 66072-66114.
8.   CPMA News Release, December 9, 1994, Color Pigments Association Opposes EPA Proposed Pigments Waste Stream Rule as Unnecessary, Unfair, and Contrary to EPA Policy.

Visual Texture

Josef Feldman
FM Group Inc., 150 Route 17, P.O. Box 46, Sloatsburg, NY 10974, USA

ABSTRACT

Surface texture imparted to polymer parts has a long history of use. These include masking surface imperfections, creating special effects, imitating natural materials such as leather or to help hide scuff marks and fingerprints. Texturing is primarily accomplished by embossing the surface of the polymer. An alternative technique for creating the visual appearance of texture while maintaining a smooth surface has been developed by creating homogeneous patterns in the polymer melt. This technique optically breaks up the surface appearance creating a visual texture. Most polymer processes including extrusion and injection molding are amenable to this technique.

THE PHYSICS OF TEXTURE

Webster s dictionary defines *texture* as "The look, surface or feel of something." We thus associate two senses: the tactile and the visual with texture, although the former clearly dominates our usage of the word. Texture is normally a pattern of deformities on the surface of a part. These deformities are on a microscale compared to the size of the part. They may be random or directional. When the scale of the deformity increases it may become a pattern.

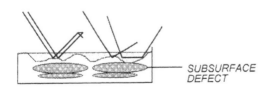

Figure 1. Light scattering.

Our binary vision system, with its great depth of field, is able to see: texture *via* the shadowing and light scattering of the deformities, Figure 1 illustrates this effect. We perceive gloss finishes as no texture and matte finish as a light scattering <u>micro-texture</u>.

PLASTIC APPLICATIONS OF TEXTURE

Surface texture is usually controlled by the properties of the surface onto which a part is cast, molded or embossed. Physical texture maybe applied to the surface to change its tactile properties. Control of slip is an important parameter in the design of some products. However most texturing of polymer surfaces seem to be used to control; "visual" effects. Texture creates a pattern of light and dark areas which might be characterized as a **shadow-scatter** mask. This mask can create a variety of useful effects. These range from surface gloss control to imparting a surface texture masking "subsurface" imperfections such as inhomogeneous melt patterns (Figure 1).

Some enduse applications involve parts that are subject to dirt or fingerprints. The breakdown of the smooth surface into a texture "shadowed" surface partially masks the dirt. Unfortunately it also contributes in the long run to "dirt trapping" which may defeat its original purpose. We need look no further than our computer keyboard to illustrate this point

VISUAL TEXTURE

Visual texture is the term we use for the technique of breaking up the appearance of a surface while leaving the physical surface smooth and unembossed.

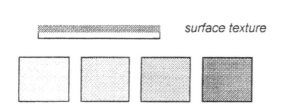

surface texture

Figure 2. Two dimensional visual texture.

Visual texturing may be accomplished by printing or painting a part with a micro pattern. This technique has recently been popularized with a variety of "granite" like appearances which have been printed and converted into melamine-laminates. Texturing paints both for imparting a physical texture as well as a visual texture are also available. These techniques suffer from a lack of "depth of image" and look flat when compared-to physical textures. They suffer a common problem with physical texture in that they are surface techniques. Abrasion of the surface will alter or destroy the effect.(Figure 2)

THROUGH-COLOR TEXTURE

To overcome these difficulties it is desirable to have a visual texture effect available as a color throughout the polymer. This can be achieved by creating polymer compounds with an inhomogeneous micropattern throughout the polymer. Depending on the translucency of the

Figure 3. Three dimensional visual texture.

part, a micropattern will be observable on the surface as well as below the surface. This combination gives a similar **shadow-scatter** effect as observed with physical surface texture, (Figure 3) thereby giving a visual impression of texture.

Physical surface texture is limited in its influence on visual appearance in that embossing causes **shadow-scatter** which affects appearance by shifting <u>lightness</u> but not <u>color</u>. The appearance adjectives of "soft" and "hard" are both associated with the surface properties: low gloss and high gloss. By creating **colored microinhomogeneities** one can in addition to varying lightness, vary the color of the micro pattern. This gives tonality to the object and imparts greater depth of image to the visual texture.

One can vary both the color and domain size of the foreground domains as well as the color of the background. Because these effects are through-color the appearance is very dependent on the opacity of the compound. A conventional colorants pigment concentration affects the perception of color intensity. Because the size of the inhomogeneous foreground domains are much larger than conventional pigments and the domain density is low, varying the concentration of the foreground domains changes the <u>perceived</u> number and size of particles thereby greatly changing the appearance. This is perhaps best illustrated by white particles in a lightly tinted black background (Figure 4). The colorant domain near the surface appears white on black; the domains or particles somewhat below the surface appear gray and of a smaller particle size. The particles deep below the surface appear deep black. The ramifications for quality control of appearance as opposed to just color are very great.

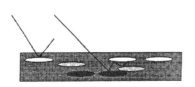

Figure 4. White foreground on black background.

CONCLUSIONS

A method of imparting a visual texture appearance throughout a polymer part has been developed. This effect is available in a variety of foreground and background colors, and domain sizes giving the appearance of a through colored texture. Applications include masking of defects, hiding dirt, simulating natural materials such as stone, marble, and leather.

Surface Smoothness and its Influence on Paint Appearance. How to Measure and Control it?

Gabriele Kigle-Boeckler
BYK-Gardner, USA

ABSTRACT

The measurement and control of color and gloss parameters is critical because of aesthetic reasons and provides an indirect measure to control process parameters. The impression gloss consists of several appearance phenomena - specular gloss, haze, DOI and surface smoothness. The evaluation of surface smoothness has major influence on the total appearance and is currently only evaluated visually. The presented new measurement technique correlates with the visual assessment of orange peel and objectively measures surface smoothness. The measurement results of the new orange peel instrument (long - and short-term waviness) were related to different process parameters, such as substrate roughness, in an actual case study.

INTRODUCTION

Especially in the automotive industry, the appearance of a painted surface becomes more and more important. An automotive finish is a multi-layer system - E-coat, Primer, Topcoat, clear coat. Each layer can have several characteristics that influence the appearance of the final finish. The two major objectives of the paint manufacturer are to optimize the physical/chemical properties of the coating as well as to achieve an optimum paint appearance (wet, smooth look). The end use of the paint (trucks versus prestige cars), then determines the necessary degree of achievement.

The coatings are usually applied to sheet metal or plastic parts. The sheet metal manufacturer has different objectives than the paint manufacturer in his production process. In the first production step, stamping, a high surface roughness is required in order to adhere a sufficient amount of grease to achieve optimum slip properties. In case of the SMC

Table 1. Combinations

Panel #	Roughness		Baking Position		Paint			Clear coat yes
	0.8 μm	1.5 μm	h	v	A	B	C	
1	x		x		x			x
2	x		x		x			
3	x		x			x		x
4	x		x			x		
5	x		x				x	x
6	x		x				x	
7	x			x			x	x
8	x			x			x	
11		x	x		x		x	x
12		x	x		x		x	
13		x	x			x		x
14		x	x			x		
15		x	x				x	x
16		x	x				x	
17		x		x			x	x
18		x		x			x	

(Sheet-Molding Compound) or SRIM (Structural Reaction Injection Molding) manufacturer, the surface quality or roughness is influenced by the type of reinforcement used (fibers, mats, roving), the particle size of the filler and the surface quality of the mold.

SRIM has been restricted to non-appearance parts because of problems with glass "read-out" and surface porosity. Recently, SRIM is used more and more for exterior automotive body panels due to its improved surface qualities. Studies have shown that a Class-A surface comparable to SMC can be obtained by selecting the proper processing parameters, reinforcements (e.g. continuous-strand glass reinforcement mat) and resins.

The paint manufacturer is now faced with the problem of a rough substrate which could have influence on the final appearance of the car finish. The sheet metal substrates are E-coated to protect the material against environmental influences (corrosion/acid rain resistance) and then a primer is applied to cover up unevenness of the sheet metal surface. In the case of SMC or SRIM parts the substrate is only primed to smooth the surface. Now the question is, how much does the roughness influence the appearance of the final finish and how can it be measured? Therefore, we have done a case study with varying parameters - roughness, baking position, paint systems and with or without clear coat - and the results on the surface appearance were interpreted with a new, optical device for measuring the surface smoothness.

SAMPLE PREPARATION

We used sheet metal with two different levels of roughness: Ra = 0.8 μm and Ra = 1.5 μm. Three different types of paint systems were applied and baked in horizontal as well as vertical position. In addition, we applied a clear coat only on half of the panels. The thickness of each layer was kept constant and the coatings were applied by the same operator. The various combinations per panel are given in Table 1.

When evaluating the panels visually the different combinations clearly showed differences in the surface appearance.

DIFFERENT TYPES OF SURFACE APPEARANCE PHENOMENA

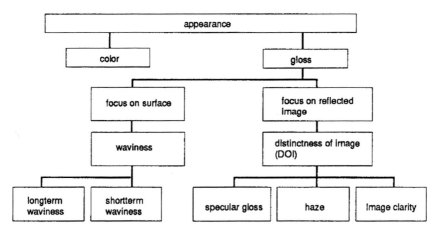

Figure 1. Different types of surface appearance.

The surface appearance is described by the color and the gloss of the sample. Gloss is the attribute of surfaces that causes them to have a shiny/mat or distinct and smooth appearance. The visual impression of gloss is derived by focusing on the surface, evaluating the contrast between highlighted and non-highlighted areas (see Fig. 2a) and by focusing on the reflected image of a light source (see Fig. 2b).

By focusing on the reflected image of a light source the distinctness of image (DOI) is evaluated. The distinctness of an image is effected by the specular gloss, haze and the image clarity (edge sharpness). A high gloss, smooth surface makes us very sensitive to haze (milky appearance). Gloss and image clarity, often referred to as DOI, are used in the automotive industry to control the paint appearance. As the quality of the paint systems has increased tremendously in the last few years, the differences in gloss and DOI became minor. However,

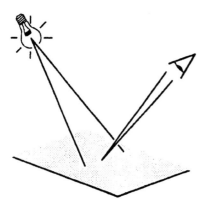

Figure 2a. Contrast.

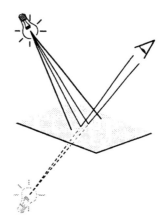

Figure 2b. Reflected image.

there are still considerable problems in controlling the overall appearance influenced by the smoothness of the surface.

When we focus on the surface, we look at the contrast between highlighted and non-highlighted areas. This type of evaluation provides us with a multitude of information about structure size, structure form and the contrast within the structure. The human eye can see structure sizes between 0.1 mm and 10 mm. Our triangulation system does not have the resolution to measure the real depths of structures on a coated surface which are between 1 μm and 2 μm. Our brain interprets the light-dark pattern that we see as a 3-dimensional structure with hills and valleys. Therefore, the contrast within a structure gives the impression of the depth of the structure and, indirectly, the smoothness of the surface is evaluated. In the automotive industry the surface smoothness is often referred to as orange peel. One can see different grades of orange peel - long term and/or short term - on a surface resulting in different appearances. Very often, a certain degree and type of waviness is intentional on a surface to cover up defects or to achieve a specific look.

In order to quantitatively evaluate the smoothness of a surface, which has considerable influence on the total gloss appearance of the sample, it is necessary to objectively measure this phenomena. The measurement of specular gloss and reflection haze does not take into account the visual impression of orange peel. Also the measurement of image clarity correlates poorly with the visual ranking of orange peel.

At the moment, the phenomena orange peel, is either evaluated visually or by profilometry, which can only be used in laboratory applications. Physical paint standards with different degrees of orange peel or other visual instruments are used as support tools for the visual evaluation (Tension Meter, Landoldt Rings). Therefore, we believed that it is necessary to develop a new method for evaluating orange peel that focuses on the surface and correlates with the visual evaluation.

ORANGE PEEL INSTRUMENT: OBJECTIVE AND QUANTITATIVE EVALUATION

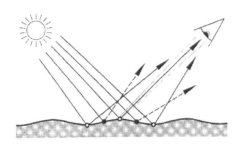

Figure 3. The visual evaluation.

Our orange peel meter has been designed to simulate the visual evaluation by focusing on the surface (see Fig. 3). The instrument uses a point source - laser - for illuminating the sample. The sample surface is scanned by moving the instrument over a distance of 10 cm. The light source illuminates the surface at 60° and the reflected light is measured at the equal but opposite angle. When the light beam hits a peak or a valley of the surface we detect a maximum signal. On the slopes a minimum signal is registered. Therefore, the measured signal frequency is equal to the double spatial frequency of the topography.

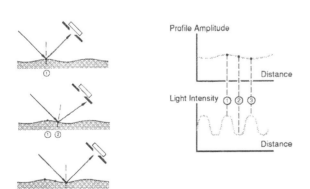

Figure 4. Measurement.

The orange peel meter measures the optical profile, "what you see", while a profilometer measures the mechanical profile, "what you feel" (see Fig. 4). In order to simulate the resolution of the human eye, we differentiated the waviness into long term and short term waviness. When visually evaluating the surface appearance, the observer usually looks at the object closely (approx. 30 cm away) as well as at a greater distance (approx. 2.5m). Being 30 cm away, the human eye can see 35 separate lines that have a width of 0.07 mm as well as a distance between each other of 0.07 mm. To simulate this situation the instrument takes a measurement every 0.08 mm resulting in 1250 measuring points per scan length (10 cm).

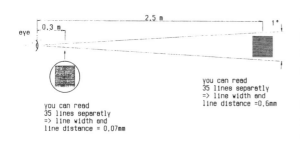

Figure 5. Visual observation.

At a distance of 2.5 m we can see 35 separate lines that have a width of 0.6 mm as well as a distance between each other of 0.6 mm (see Fig. 5). To simulate this situation the raw data are divided into long wave (structure size > 0.6 mm) and shortwave (structure size < 0.6 mm) signals by using a mathematical filter function. The long term waviness represents the variance of the long wave signal amplitude and the short term waviness represents the variance of the shortwave signal amplitude.

In addition, long and short term waviness values were mathematically correlated to other industry known, visual measurement scales (tension value, orange peel standard ranking) and are calculated by the instrument. The correlation to visual scales was implemented to facilitate the evaluation, but at the same time, more detailed information for interpreting the cause of the surface appearance (long wave, shortwave signal) are displayed.

RESULTS OF EXPERIMENT BY USING THE ORANGE PEEL METER

The questions to be answered are: "What is the effect of roughness, paint system, baking position and additional clear coat on the long term and short term waviness values?"

These questions can be answered by sorting the measurement results by long term waviness and short term waviness.

SHORT TERM WAVINESS RESULTS

The results are included in Table 2 and Figure 6. The panels with the lower level of roughness and the additional clear coat had the lowest short term waviness values. The group with the second best results had the lower level of roughness, but no clear coat layer. Next were the panels with higher roughness and clear coat. The worst results were measured on the panels with the higher roughness and no additional clear coat. Therefore, we can conclude that the changes in short term waviness are caused by the substrate roughness and the additional clear coat layer. The baking position and the paint system had hardly any influence on short term waviness.

LONG TERM WAVINESS RESULTS

The results are included in Table 3 and Figure 7. The panels baked in the horizontal position with a lower level of roughness had excellent results for long term waviness. The second best

Table 2. Results

Panel #	Roughness		Baking position		Paint			Clear coat yes	Long wave	Short wave
	0.8 μm	1.5 μm	h	v	A	B	C			
1	x		x		x			x	0.9	0.9
3	x		x			x		x	1	1.7
5	x		x				x	x	1	1.9
7	x			x			x	x	11.6	3.8
4	x		x			x			0.7	5.5
2	x		x		x				1	8.5
6	x		x				x		1.5	8.7
8	x			x			x		6.5	9.4
15		x	x				x	x	1.5	9.6
11		x	x		x			x	1.8	13.3
13		x	x			x		x	2.5	15.3
17		x		x			x	x	25	18.1
18		x		x			x		16.2	29.5
16		x	x				x		5.4	31.1
14		x	x			x			3.8	32.9
12		x	x		x				3.9	33.9

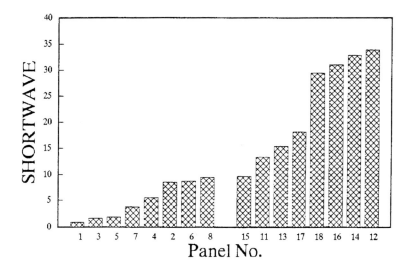

Figure 6. Short term waviness results.

Table 3. Results

Panel #	Roughness 0.8 μm	1.5 μm	Baking position h	v	Paint A	B	C	Clear coat yes	Long wave	Short wave
4	x		x			x		x	0.7	5.5
1	x		x		x			x	0.9	0.9
3	x		x			x		x	1	1.7
5	x		x				x	x	1	1.9
2	x		x		x				1	8.5
6	x		x				x		1.5	8.7
15		x	x				x	x	1.5	9.6
11		x	x		x			x	1.8	13.3
13		x	x			x		x	2.5	15.3
14		x	x			x			3.8	32.9
12		x	x		x				3.9	33.9
16		x	x				x		5.4	31.1
8	x			x			x		6.5	9.4
7	x			x			x	x	11.6	3.8
18		x		x			x		16.2	29.5
17		x		x			x	x	25	18.1

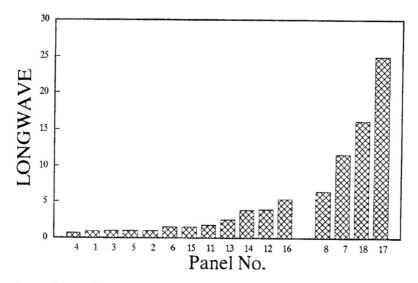

Figure 7. Long term waviness results.

results were obtained on the panels baked in the horizontal position and the higher level of roughness. The worst results were recorded on the panels baked in a vertical position with a high level of roughness. That means, the changes in long term waviness are affected mainly by the baking position and secondly by the roughness and the additional clear coat. The paint system had no significant influence.

SUMMARY

In this paper the different parameters of gloss and their effects on the paint appearance were discussed. The visual impression of gloss is a combination of focusing on the reflected image of a light source and focusing on the surface, evaluating the contrast between highlighted and non-highlighted areas. In a case study the influence of substrate roughness, baking position and clear coat layer on the surface appearance are evaluated by using a new objective instrumental method for measuring orange peel. To evaluate the appearance phenomena, orange peel, the instrument simulates the human eye by focusing on the surface and measuring the contrast between the dark and light areas. The measurement results are divided into long term and short term waviness which have a direct correlation to different process parameters such as substrate roughness, baking position and with or without clear coat.

REFERENCES

1. Kristen L. Parks, Randall C. Rains, *Class-A Structural: Report on a Program to Develop SRIM for Exterior Automotive Body Panels*, SAE Paper No. 920371, 1992.
2. Horst Schene, **Untersuchungen ueber den optisch - physiologischen Eindruck der Oberflaechenstruktur von Lackfilmen**, *Springer Verlag*, Berlin, Heidelberg, 1990.
3. Monica H. Verma, Gregory W. Cermak, *Orange Peel and Paint Appeal,* ISCC Williamsburg Conference on Appearance, 1987.

Static Control Methods in Plastics Decorating to Reduce Rejection Rates and Increase Production Efficiency

H. Sean Fremon
Meech Static Eliminators, USA

ABSTRACT

Untreated static electricity in plastic decorating operations can result in dust contamination, high scrap and rework rates, production slowdowns, improper laydown of inks and coatings. When static is treated correctly, the results can be astounding, measured in profit rates changing and increased speeds of production lines. We will cover many of the decorating processes plastics processors incorporate, and show how to control the related static and dust problems routinely seen in painting, coating, screen and pad printing, assembly and hot stamping.

BACKGROUND

In high quality decorating and coating operations, the costs of untreated static can be tremendous. Some plants expect scrap rates to be around 10 percent and price their products accordingly. Much, but not necessarily all, of that 10 percent can directly be attributed to static, and, if the proper equipment is employed, scrap rates can be significantly lowered. Nobody who decorates plastic is immune to static: bottle printers, painters of automotive parts, hot stampers of toothbrushes, pad printers of cellular phone buttons or coaters of lenses, they all can be affected.

In a typical plant, people play the blame game: decorators blame molders and molders do the best they can to keep the products clean. However, all of the processes incorporated into manufacturing and decorating plastic products leave lots of static and its related problems.

People in the metal industries do not have these problems. But, you may wonder, what exactly is static? Well, let's define it. Static is the excess or deficit of electrons on a material. A

The Triboelectric Series is a sequence of common materials and substrates that are arranged in such a way that any one of them is positively electrified by rubbing it against another substance farther on the list. For example, if you were to rub polyester against an acetate fiber, the acetate fiber would gain the positive charge and the polyester the negative. But the polyester would gain a positive charge if rubbed against cling film, PVC or Teflon; likewise the acetate fiber would gain a negative charge if rubbed against any substance higher up on the list. Note that many of these materials are often encountered during plastics decorating and assembly.

Positive charge

Air
Human Skin
Rabbit fur
Glass
Human hair
Nylon
Wool
Silk
Aluminum
Paper
Cotton
Steel (neutral)
Wood
Hard rubber
Nickel, copper
Brass, silver
Gold, platinum
Acetate fiber (rayon)
Polyester
Cling film
Polyethylene
PVC
Teflon

Negative charge

Figure 1. Triboelectric series.

regular nonpolar atom has a nucleus with a positive charge and negatively charged electrons orbiting it. When one atom comes in close proximity to another, electrons from the orbit of one atom draw or tear away electrons from the other, leaving one atom with an excess of electrons and the other with a deficit. This process, called ionization, leaves the atom with the excess carrying a negative charge, while the one with the deficit carries a positive charge.

Why does static make products behave like they do and cause quality problems? Most of it has to do with opposite charges attracting and like charges repelling. Static in plastic decorating operations are most often caused by friction, separation, heat change and improper grounding. Normally a processed plastic part experiences all of these factors.

SOURCES OF STATIC ELECTRICITY

Friction. Static is most commonly caused by friction, the rubbing of two materials together during the molding or product handling process. When plastic resin rubs against a metal vacuum tube, both become charged. The plastic picks up the negative charge while the metal tubing picks up the positive charge. Because metal is such a good conductor, however, the charge on the metal bleeds off quickly. The charge on the plastic does not.

If you have ever wondered whether a material is most likely to charge positively or negatively when rubbed against another, the Triboelectric Series (Figure 1) will generally tell you. The materials higher on the chart will hold the positive charge, while those lower on the chart will hold the negative charge.

Separation. If you have ever unrolled stretch wrap in your factory, and the hair on the back of your arms is raised, this is the result of static. This is also the case when a part is removed from its tooling after molding, or a screen is drawn away from the part it just printed.

Once full of static the part begins to attract airborne dust and particles of the opposite charge. Here is an example: If you have a piece of Scotch tape, pull off a piece (separate it from its roll) and hold it about an inch off the floor. It will act as a magnet, pulling all different types of dust from the floor. This same effect happens on a much larger scale with larger statically-charged plastic products, which can sometimes pull airborne contaminants from more than 4 feet away.

Heat change. Continuous heat change can build a static charge on many products. Dryers cause this phenomenon constantly. If you have a multiple coating process, the constant heating and cooling of the part can lead to the product becoming highly charged and contaminated. **As a part cools its charge increases.** If you have come in contact with a dryer conveyer belt after a screen printed plastic bottle has been dried, chances are good that you have received a shock. Because charges and temperatures are changing during the drying phase, dust can be drawn to the area. In extreme situations, the dust can be drawn to the product surface and adhere to the product making it nearly impossible to remove later.

Improper grounding. One of the most common misconceptions in plastic decorating is that if a part is grounded, it must be static free. However, despite equipment grounding efforts, greased bearings and poor metal-to-metal connections can interrupt the ground, and some charges may be too high to ground properly. Because plastic parts (and many of the inks and coating used to decorate them) are insulators, you can partially reduce static through grounding but you can never fully eliminate it.

MEASURING STATIC ELECTRICITY

Static electricity is measured in kilovolts and can range from 0-200 kV and even higher in decorating and assembly operations. At about 30-40 kV, operators can receive nasty shocks. At 5 kV, static begins to draw airborne contaminants to a product. Static meters can be an essential tool in assuring that a problem thought to be static really is static-related, and not coating or ink related for example. Most larger manufacturing facilities have these devices.

RESULTS OF UNTREATED STATIC ELECTRICITY

Dirt. Decorators of plastic know all too well the nuisance of dirt and contamination. "Dirt" can include dust generated from industry, sawdust and drilling dust, lint or fabric, reground plastic, corrugated box dust, paint flakes and a variety of other materials. The statically-charged contaminants attach themselves to other products of opposite charges, making a static bond between the two. The product is then decorated, coated or printed and appears as a defect, in some cases even looking larger than it actually is. However, if proper static controls are in place these problems can be alleviated.

Image spiders and webbing. One of the most common quality issues, for which static is at least partially responsible for, is ink "spidering" or webbing. After the squeegee is rubbed against the screen repeatedly or the tampon prints for a while, the friction causes a charge to build up in the polyester mesh. The screen, normally negatively charged, is forced onto a plastic substrate, also negatively charged, in the printing cycle. The two negative charges repel each other, causing the ink to misbehave and creating a mark similar to a lightning bolt on the image. In one operation, it was reported that the proper use of static control led to a large quality increase. Where static control was employed, the operator could produce an image of 120 dpi, while without static control it could only be run at 85 dpi.

Operator shocks and damage to and **interference in electronic components.** Strong static fields and static discharges can cause unpleasant operator shocks and interfere with electronic components. Concerns arise when the operator recoils from the shock near moving machinery. Equipment can also fail as a result of high charges. Ink jet printers and labeling equipment is susceptible to static damage because of contact grounding of the product.

TYPES OF STATIC CONTROL DEVICES

Static control devices are made in many different shapes and sizes, so they can be retrofitted to different applications of several different products. They can be classified into four different groups: (1) Active AC electrical eliminators, (2) Active electrical DC eliminators, (3) Nuclear, and (4) Passive or grounding devices. Some incorporate air to carry the ions to the part while others do not.

AC electrical eliminators. Active electrical AC ionizing equipment, one of the most common and effective ways of eliminating static charges, is seen in a large number of plastic decorating operations and assembly areas. These devices, which account for the highest volume of static control used in industry, are custom manufactured in different shapes and sizes. They are adaptable to most static situations and retrofitable to the machines they serve. Most, without the use of air, are only effective within a few inches of the target. They include bars and rods, ionizing fan-driven blowers, air curtains and amplifiers, nozzles and guns.

The units create ionization by forcing high voltage to ground and breaking down the air molecules in between into low levels of ozone. Such ionizers make available millions of free positively and negatively charged ions and effectively "feed" away the static charge by "matching up" ions of opposite charges.

Electrical ionizers are driven by 4500-9000 volt power supplies which transform the voltage from 110 v or 240 V, while lowering amperages to a safe level. If this equipment is not employed properly, in many situations it is useless.

DC ionizers. DC ionizers, or cloud ionizers, are employed in areas where air cannot be used as an ion carrier such as a cleanroom. This makes them ideal in decorating because they can be used in screen or pad printing areas to discharge the screen, but not dry the ink. Pulsed DC controllers allow adjustment of positive and negative ions and control of the frequency of ions being emitted (measured in hertz).

Nuclear ionizers. Nuclear ionizers, or radioactive ionizing devices, are seen in plastic decorating operations where solvent-based inks and coatings are used. Electricity cannot be used near because the possibility of an electrical spark exists. Nuclear ionizers incorporate Polonium-210, an alpha-emitting isotope. When the alpha particles collide with an air molecule, the collision creates thousands of positive and negative ions without the use of electricity. Most nuclear ionizing units have an active life of approximately one year, after which they must be returned and refilled.

Passive systems. Passive static control incorporates grounding and conductive materials. Gold and silver are excellent conductors but are cost prohibitive. Carbon fiber, stainless steel and phosphorus bronze are commonly used passive static-control devices in plastic processes and decorating. However, copper tinsel remains the most common because of its low cost and availability. Its efficiency is considered mixed; but if it reduces static to an acceptable level, it does away with the need for more expensive electrical ionizers or controls.

USES OF STATIC CONTROL DEVICES IN DECORATING AND ASSEMBLY

Several methods of static control have been successful in different facets of decorating and assembly. In some parts of the manufacturing processes it is better to try to prevent dust attracting throughout the process than after the part has been contaminated. In others it is more logical to remove the dust inline because of the plant layout or potential for contamination throughout molding and secondary processes.

IN-LINE PAINTING

Dirt remains the biggest concern among plastic decorators as a problem resulting in rejects or timely reworks. Once the product is contaminated, two options remain in cleaning it so that it can be decorated: power washing with water or part blow off with ionizing units. While any part may only be reworked 2-3 times before it is scrapped, it is important (and less expensive) to have the product clean during its first pass.

Cleanliness is most important for people who are molding and painting plastic parts where quality is most important. Automotive parts must look better than garbage cans, for example. Many manufacturing plants report scrap rates of 10 percent or more due to dirt, paint overspray, bad paint mixes, color variations, runs and other processing mishaps. There are three ways dust and contaminants can be blown off. They are (1) Ionizing air guns, (2) Ionizing air curtains, (3) Self-contained fan-driven units.

Ionizing Air Guns. Ionizing air guns are used for cleaning small plastic parts and in areas where touch up is necessary before painting. While they are time consuming for an operator to use cleaning large parts, such as cars, instrument panels or fascias, they are widely used for smaller parts. Only a compressed air source is needed and electrical power for electrical guns to operate.

Ionizing air curtains. Also called air knives, these units also use compressed air and are used in many automated painting operations. Most often they are hung vertically or in a halo around the conveyed plastic product. An air curtain is constructed so that compressed air is brought in through the end and exits through a narrow, elongated slot in the device. As the air leaves the air curtain, it passes over an ionizing bar to carry free ions to the product it is blowing off. It also entrains ambient air to drive it to the product, amplifying the air up to 20 times, while lowering the draw from the compressor. While this process of dust control is moderately effective, it also has its down sides because the entrained air may be dirty or contaminated with dust and be driven into the product before painting.

These pieces of equipment are also expensive to operate. With the industry standard price of compressed air at approximately 29 cents/1,000 cfm, oftentimes a new compressor is needed or the costs do not justify using the ionizers. At one company, the cost for operating compressed air ionizing units on a single line was $200,000 per year.

High Volume Fan Driven Blower. The most economical way to reduce dust for larger parts prior to painting in many plants is by using a high-volume fan driven blower with ionizing manifolds that do not entrain ambient air. Higher air volumes and velocities are attainable and dust is not blown toward the part. It is standard at the intake of the blower to have a 10-micron filter to keep dust and contaminants out of the system. Depending on the overspray and paint collection system, it may be necessary to have a dust collection system to remove the excess dirt. Oftentimes when the static is taken out of dirt, gravity will pull it to the ground. Some plants have reported drops of scrap from six percent to less than one percent.

SCREEN AND PAD PRINTING

In screen printing and pad printing, untreated static electricity can result in image spider and webbing, dirt and dust on the product, tampon or screen, and shocks to your people.

Proper static control can remedy this problem. A long-range static eliminator placed above the screen or pad printing area will flood the area with millions of free positive and negative ions, eliminating the charge on both the screen and the pad and the substrate. Since ions are smaller than screen mesh, the right type of static control equipment is capable of neutralizing static on both the mesh and substrate simultaneously.

ASSEMBLY

Most static problems in assembly are related to automated assembly lines and feeder bowls. The most effective means of static control is using DC or cloud ionization over feeder bowls. The parts in the feeder bowl, while statically charged, will tend to stick to one another due to the vibration, and not feed properly. This is especially noticeable in small parts like pens, or components of medical devices.

In hand assembly of electronics or medical devices, cleanliness is critical. The use of pulsed DC technology at benchtops is common, and very effective in lowering part contamination levels.

TAKING CHARGE OF STATIC PROBLEMS

Static control devices, when used properly, can make a great deal of difference in decorating quality, production efficiency, overall costs and timely reworks or scrap. It is important that you know your application well when you are specifying or purchasing static-control devices to ensure that the equipment is necessary and correct for the application for which it is intended. If you are not sure, you should meet face-to-face with your static-eliminator supplier. When ionizer results are minimal, it is usually because equipment is improperly placed or the wrong equipment is being used. But if the right equipment is used for the right application, the results can be huge savings in downtime, quality control and production quality.

Dispersive Mixing of Surfactant-Modified Titanium Dioxide Agglomerates into High Density Polyethylene

Javier P. Arrizón, Rafael E. Salazar and Martin Arellano
Departamento de Ingeniería Química, Universidad de Guadalajara Blvd., Gral Marcelino Garca Barragan # 1451, Guadalajara, Jalisco C.P. 44430, Mexico

ABSTRACT

The description of the dispersion process during the manufacture of composites is an important factor in designing polymer processing machines. In this work, the mixing of titanium dioxide agglomerates into high density polyethylene (HDPE) has been studied. The surface of the titanium dioxide particles was treated with an anionic surfactant to compatibilize this polar powder with non-polar polymers, such as HDPE. The mixing process was carried out in an internal mixer equipped with roller blades at different operational conditions and using various compounding sequences. The rheological properties for these filler-polymer systems were measured using a dynamic rheometer. The energy requirement for dispersive mixing was obtained from the torque-time curves. The surface modification of the titania particles improved the dispersibility of the powder into this non-polar hydrophobic polymer.

INTRODUCTION

The degree of mixing of additives such as plasticizers, stabilizers, pigments, fillers and reinforcing materials is one of the most important factors affecting the quality of a finished plastic product. Most of the additives tend to agglomerate, which makes difficult the dispersion process. Manas[1] gives an excellent review on the dispersive mixing of solid additives. In that paper the dispersion process is divided into several stages: incorporation of the solid, the wetting of the particles by the polymer melt, the breaking-up of the agglomerates by the hydrodynamic shear stress and the distribution of the fragments in the melt. In order to transfer enough shear stress to overcome the cohesive forces in the agglomerate, good wetting of the particles by the melt is a required condition. To improve the dispersibility of such additives, it is necessary to modify the surface characteristics of the particles by a treatment.

Titanium dioxide is the most used white pigment in the plastic industry. This pigment is hydrophilic and the dispersibility in non-polar polymers such as polyethylene is not easy. In the previous work, the adsorption of anionic surfactant has been used to modify the hydrophilicity of titania.[2-4] It was found that the surfactant treatment improves the dispersibility of titania in non-polar media.

EXPERIMENTATION

The materials used in this work were titanium dioxide (TiO_2) from DuPont (commercial grade), Aerosol OT (anionic surfactant) from Aldrich Co. and high density polyethylene PEAD-65050 (melt flow index 5 g/10 min) from PEMEX. The titanium dioxide particles have an inorganic treatment (alumina 5 % and silica 3%). The pigment particles underwent two different treatments: a) the powder was screened and dried under vacuum at 100°C for 24 hrs; b) a coated powder was prepared by adsorbing the surfactant from an aqueous solution. This procedure is described elsewhere.[5]

The dispersion experiments were carried out in an internal mixer equipped with roller blades (Haake, Rheomix 600). Three different compounding sequences were used:

A) In this mixing procedure, the titanium dioxide powder and the polymer pellets were charged into the mixer chamber simultaneously and the rotors started during this process.

B) In this technique, the 70% of the polymer was first charged into the chamber and melted. Then, the TiO_2 is added at very low rotational speed (4 rpm) to incorporate the agglomerates into the polymeric melt. After that, the rest of the polymer was charged with the rotors moving. Finally, the rotors were stopped and then restarted at high rotational speed.

C) The third dispersing method is similar to the one described in part A, but in this case artificial agglomerates were used. The artificial agglomerates were made by compaction of the powder to a given density. The compact was broken to small fragments and screened to obtain agglomerates smaller than 1 mm.

In the above experiments, operational conditions such as the rotor speed (20-40 rpm) and temperature (170-200°C) were varied.

For the rheological experiments, samples from the first compounding method were molded to rectangular sheet using a press. The rheological data were obtained using a dynamic rheometer (Rheometrics RDS-II) using a parallel plate geometry. The experiments were carried out at 10% strain and gap of 1.0 mm.

RESULTS AND DISCUSSION

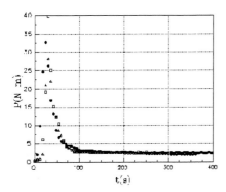

Figure 1. Torque-time curve for dispersion process using technique A (T=170°C and 20 rpm). • pure HDPE, □ HDPE-TiO₂, Δ HDPE-TiO₂/AOT.

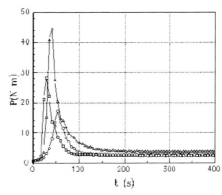

Figure 2. Torque-time curve for dispersion process using technique A (samples with surfactant treatment, T=170°C and 20 rpm). Δ 50%, ○ 30%, □ 15%.

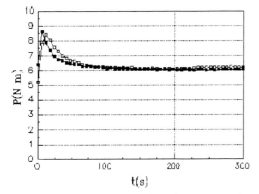

Figure 3. Torque-time curve for dispersion process using technique B (T-170°C and 20 rpm). ■ 50% TiO₂, □ 50% TiO₂/AOT.

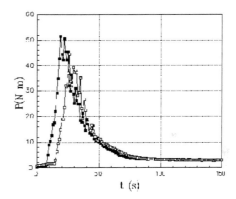

Figure 4. Torque-time curve for dispersion process of artificial agglomerates using technique A (ρ_aggl=1.61 g/cm³, T=170°C, and 20 rpm). ■ 30% TiO₂, □ 30%TiO₂/AOT.

Figure 1 shows a typical torque (P)- time curve for the pure polymer and for the TiO₂-HDPE mixtures using the technique A. It can be observed that the torque value at the peak for the bare powders is larger than the one for the treated pigment. Moreover, the equilibrium value is similar for both systems and higher than the pure polyethylene. Even though the cohesivity of the AOT treated agglomerates is higher than the bare titania according to the tap density of the powders (ρ_{TiO2}= 1.1 g/cm³ and $\rho_{TiO2/AOT}$= 1.4 g/cm³) the dispersibility of this powder is better.

The effect of the composition for the surfactant treated powders is shown in Figure 2. It can be observed that the peak and the equilibrium values increase with concentration of powder.

The torque - time curve for the technique (B) is presented in Figure 3. It can be observed higher torque at the peak for the non-treated powder.

In Figure 4 the torque curves for the dispersion of artificial agglomerates is shown. As in the case of non-compacted agglomerates, the surfactant treated powder requires lower torque value compared with the bare powder.

As a dispersion index,[6] the energy (ED) required to produce a steady-state dispersion in the matrix-pigment system was evaluated using equation [1]

$$ED = 2\pi N \int_{t_o}^{t_e} P(t)dt \qquad\qquad [1]$$

where N is the rotational speed in revolutions per second, t_o marks the time when the dispersion process starts, t_e is the time to reach the equilibrium state. The dispersion energy for all the experiments is summarized in Tables 1 and 2. In all the cases the surfactant treated TiO_2 requires less energy.

Table 1. Summary of the dispersion experiments by technique A (T=170°C and 20 rpm)

System	Dispersion energy, J			Equilibrium torque, Nm			Rate constant, k, s^{-1}		
System	15%	30%	50%	15%	30%	50%	15%	30%	50%
HDPE/TiO$_2$	2441.1	2520.1	3987.5	2.44	2.77	4.20	2.76	1.93	1.50
HDPE/TiO$_2$/AOT	1829.6	2223.3	3779.5	2.43	2.72	3.89	3.44	1.85	1.12

Table 2. Summary of the dispersion of artificial agglomerates (ρ_{aggl}=1.61 g/cm^3, T=170°C, and 20 rpm)

System	Dispersion energy, J	Equilibrium torque, Nm	Rate constant, k, s^{-1}
HDPE/TiO$_2$	3113.3	2.80	1.41
HDPE/TiO$_2$/AOT	2773.2	2.78	1.72

The kinetics of the dispersion process was analyzed according to the model presented by Cotten.[7] The torque was transformed to a dimensionless variable (P*) using

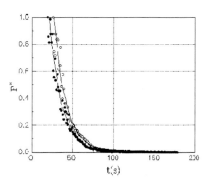

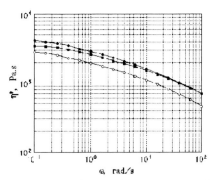

Figure 5. Dimensionless torque-time curves for dispersion process of artificial agglomerates using technique A (ρ_{aggl}=1.61 g/cm³, T=170°C, and 20 rpm). ○ 30% TiO₂, • 30% TiO₂/AOT, solid lines represent the model.

Figure 6. Viscosity as function of frequency at T=200°C. ○ pure HDPE, ■ HDPE-30% TiO₂, • HDPE-30% TiO₂/AOT.

$$P^* = \frac{P - P_\infty}{P_0 - P_\infty} \qquad [2]$$

where P_0 is the torque at the peak, P_∞ is the torque for long mixing times (steady-state value). The dimensionless torque can be described by an exponential decay

$$P^* = \exp(-kt) \qquad [3]$$

where k is the rate constant.

From the fitting of the experimental data to the equation [3], the rate constant for this process can be obtained. In Figure 5, the experimental data are compared to the model. The agreement between them can be observed. The rate constants are summarized in Tables 1 and 2. The k values decrease with the increment of TiO₂ concentration. For the case of compacted agglomerates, the surfactant treated powders are dispersed faster than the non-treated powders.

Finally, the rheological behavior of the pure and the filled polymer at different temperatures are presented in Figure 6. As expected the complex viscosity of the melt is increased by the addition of the titania particles. The coated powders showed a smaller increment in the viscosity, which can be translated to better processability.

CONCLUSIONS

The use of surfactant to modify the surface characteristics of hydrophilic powders has been found to be an effective alternative to improve the dispersibility and processability in non-polar polymers. The simple model proposed by Cotten described with excellent agreement the dispersion kinetics.

REFERENCES

1   I. Manas-Zloczower, **Mixing and Compounding of Polymers: Theory and Practice**, I Manas-Zloczower, and Z. Tadmor, Eds., Chap. 3, *Hanser Publishers,* New York (1994).
2   I. Manas-Zloczower, D.L. Feke, M. Arellano, Y-J. Lee, *J. Macromolecular Science. Pure and Applied Chemistry,* **A32** (8&9), 1621(1995).
3   M. Arellano, I. Manas-Zloczower, D.L. Feke, *Polymer Composites*, **16** (6), 489 (1995).
4   M. Arellano, I . Manas-Zloczower, D.L. Feke, *J. Coating Technology*, **68** (857), 103 (1996).
5   M. Arellano, **Ph.D. Thesis**, *Case Western Reserve University* (1995).
6   M. Y. Boluk, H.P. Schreiber, *Polymer Composites*, **10** (4), 215 (1989).
7   G. R. Cotten, *Rubber Chemistry Tech.*, **57**, 118 (1984).

A Comparative Study of the Use of High Intensity Dispersive Mixers and Co-Rotating Twin Screw Extruders in the Manufacture of High Quality Color Concentrates

Alex Rom-Roginski
Colortech, Inc., Morristown TN 37814, USA
Christopher J. B. Dobbin
Industrial Research+Development Institute, Midland, ON L4R 4L3, Canada

ABSTRACT

Over the past three decades, high intensity dispersive mixing (HIDM) technologies have been adapted for specialized applications in the polyolefin and PVC compounding industries. High intensity dispersive mixers (also referred to as thermo-kinetic mixers) are unique in their ability to rapidly mix and melt polymer preparations in a single high-speed operation. This study was undertaken to provide a direct comparison of commercial-scale HIDM and twin screw extrusion compounding technologies for the preparation of polyolefin color concentrates. Five commercial concentrate formulations were prepared independently on a 40 liter HIDM and a 60 mm co-rotating twin-screw extruder. A detailed analysis of dispersion quality and color strength is presented. Results indicate that these compounding technologies can provide virtually identical product quality when operating conditions are properly optimized.

INTRODUCTION

CONVENTIONAL COMPOUNDING TECHNOLOGIES

Although the scope and demands of polymer compounding have increased dramatically over the past 30 years, the nature of the equipment used by the compounding industry have remained largely unchanged. With few exceptions, the primary compounding tools are comprised of extruders (single screw, counter-rotating, co-rotating and non-intermeshing twin screws), continuous mixers and low speed, high-intensity batch mixers. In their various forms, they all attempt to melt, mix, devolatize and pressurize the polymer melt while simultaneously optimizing filler and pigment dispersion, throughput rate and product consistency.

The selection of a specific processing technology is highly dependent on the nature of the materials being processed. In the preparation of polyolefin-based color concentrates, continuous mixers are generally considered to exhibit superior control of inorganic pigment dispersion over a broad range of loading levels. They typically find application in the preparation of highly loaded 50-80% titanium dioxide white formulations, for example. Intermeshing twin screw technologies are preferred for difficult to disperse organic pigment systems or unusually slippery additive systems.[1] The modular nature of the twin screw extruder allows a high level of selectivity when it comes to location (and intensity) of mixing elements and the placement of vacuum or atmospheric venting ports and supplementary downstream feeders. Most dispersion problems (and many high lubricity additive situations) can be adequately dealt with, although throughput rates are often sacrificed. In both systems, the commercial requirements of high throughput have a negative impact on overall dispersion quality due to the high forces brought to bear on the pigment particles during the resin melting stage. In the later 1970's and early 1980's, a number of alternate compounding technologies were developed in an attempt to more effectively integrate all (or most) of the critical compounding functions in a single device. Devices such as the Instamelt rotary extrusion system[2] and the Farrel Diskpack[3,4] achieved significant technical success but were largely commercial failures. At about this time, Colortech began to explore the potential of the high speed, high intensity thermo-kinetic melt mixer for the production of polyolefin color concentrates. By 1996, Colortech was producing more than 15 million kg per year of high quality concentrate product using thermo-kinetic mixing technologies exclusively.

The recent acquisition of Colortech by PPM, a German color compounder, presented an opportunity to conduct a direct comparison of thermo-kinetic and conventional twin screw compounding technologies. The goal was to run identical color concentrate formulations on commercial scale processing equipment operated by individuals skilled in the art. Primary performance criteria were to include dispersion quality and color strength. Secondary characteristics such as melt index, ash content and product density were also evaluated.

HIGH INTENSITY DISPERSIVE MIXER - OVERVIEW

The Colortech thermo-kinetic processing system, dubbed the HIDM (for High Speed Dispersive Mixer), is similar in function to the Gelimat (Draiswerke, Inc.)[5] and the K-Mixer (Synergistics Industries Limited).[6] Although these systems differ in a number of specific design refinements and control strategies, they follow the same basic principles of operation.[7-9]

The HIDM is designed to rapidly mix and melt polymeric materials but does not pressurize the resulting melt, relying instead on specialized downstream equipment for this opera-

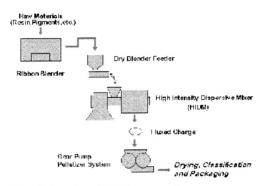

Figure 1. Overview of HIDM compounding process.

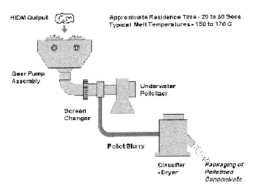

Figure 3. Detail of HIDM gear pump assembly.

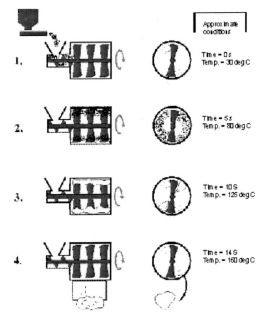

Figure 2. HIDM dispersion/fluxing mechanism.

tion. Studies suggest that a drag flow model[10,11] can approximate the primary mechanism for melt mixing.

The device itself is comprised of a water-cooled cylindrical chamber attached to feed barrel (Figures 1, 2, 3). The central drive shaft is coupled directly to a powerful, high-speed electric motor. The shaft is fitted with an integral feed screw and a series of staggered mixing blades. The number, position and pitch of the blades are optimized to transfer kinetic energy rapidly and efficiently and can be arranged to provide a self-wiping action to the inside walls of the chamber. The drive shaft is continuously rotating at speeds sufficient to generate angular tip velocities of 35-45 m s^{-1}. The clearance between the blade tips and the chamber wall is typically between 5 and 10 mm, resulting in only low to moderate shear fields being generated during the fluxing phase.

In operation, a single weighed charge of pre-blended polymer resin, pigments, fillers and additives is introduced directly into the feed throat of the HIDM. Alternately, each blend component may be gravimetrically dispensed onto a slide gate, which is then opened to drop the materials onto the rotating feed screw. The charge is transported rapidly (<0.5 s) into the mixing chamber where the particles form a "fluidized" suspension due to the action of the mixing paddles. Concurrently, impact with the rotating paddles transfers kinetic energy to the suspended materials, quickly raising their mean temperature. The intensive mixing action at this stage also breaks apart pigment and filler agglomerates.

As the temperature increases, the resin particulates begin to soften and individual pigment and filler particles begin to embed themselves in the polymer surface. As the temperature continues to climb, the polymer resin forms a molten mass that continues to turn in the mixer. This is termed "the flux point." The flux point can be readily detected by monitoring the step-wise increase in amperage drawn by the drive system. The material is kept in the mixing chamber for an additional period of time to further increase the temperature of the melt for downstream processing. Once the material has reached the desired melt state, a door in the chamber wall is opened and the material is discharged by centripetal force (usually as a single piece). The melt may then be fed into a gear pump assembly[12] or a wide throated extruder for pressurization and pelletization. Alternately, the molten material can be passed on to a downstream forming process such as calendering or compression molding.

A 40 liter HIDM production line running polyolefin concentrates can typically process shots weighing 6 to 17 kg with a flux time of less than 10 seconds and a total time-to-discharge of less than 15 seconds per shot. The dusty nature of the process is thought to be the limiting factor in achieving critical levels of pigment dispersion.

HIDM OPERATIONAL PARAMETERS

The mass of the raw material charge introduced to the mixer has a significant effect on the speed and efficiency of the fluxing operation[8] but the exact relationship between charge size, composition, dispersion quality and flux time are complex and poorly understood. As with other compounding technologies, there are tradeoffs between throughput rate, dispersion quality and product consistency. In general, larger charge sizes provide a higher density of "fluidized" material in the mixing chamber during the early stages of processing. The rate of kinetic energy transfer from the rotating paddles is also high, resulting in extremely rapid thermal heating. Discharge times are very short thereby maximizing production rates. Accurate control of the discharge point becomes critical, however, when materials are heating at a rate of 10 to 15°C per second. Poor control of the discharge state can lead to "hot shots" - a situation that can compromise product quality, stability, downstream handling and the safety of manufacturing personnel.

At the other extreme, small charges produce very low material densities within the HIDM mixing chamber, reducing the overall effectiveness of energy transfer. This can result in long mixing times prior to melting. In certain instances, extension of the high intensity mixing phase can provide additional time to break down pigment agglomerates and aggre-

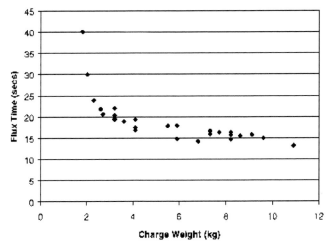

Figure 4. Effect of charge weight on flux time for granular LLDPE processed in a 40 liter HIDM.

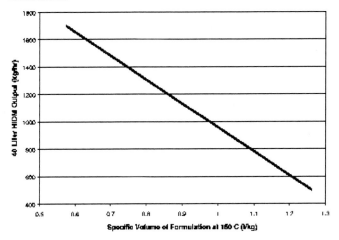

Figure 5. Average output rates as function of specific volume for 40 liter HIDM at tips speeds of 40-42 m s^{-1}.

gates, improving the overall quality of the final product. In other pigment systems, additional high intensity mixing has little positive effect and in some cases may actually be detrimental to dispersion quality. In any event, the cost of smaller charge size is lowered production rates.

If the charge size is too small, the flux time can become so long that the composition of the material in the chamber actually changes, typically by loss of low density or small particle size components as dust. In the extreme case, the amount of thermal energy generated by the paddles moving through the material is less than its heat loss to the internal mixer environment. The charge never gains enough net thermal energy to soften and melt.

Experience has shown that optimum pigment and filler dispersion result when the specific volume of the material charged to the mixer falls within a well-defined range (see Figure 4). The specific volume of the formulation is the volume (in liters) occupied by 1.0 kg of molten compound at 150°C. For a mixer with an approximate free volume of 40 liters, optimum results are obtained using specific volumes of 4.2 to 5.0 liters (10.5-12.5% of mixer free volume). Since specific volume of the melt is directly related to the product formulation, output rates vary volumetrically as a function of product density. Idealized output rates for a 40 liter HIDM are shown in Figure 5.

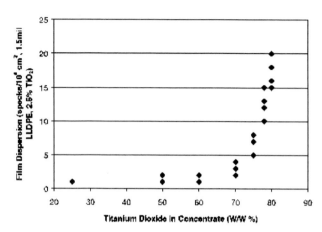

Figure 6. Dispersion quality of titanium dioxide pigment concentrates as measured in LLDPE blown film. Letdown levels have been adjusted to give 2.5% pigment.

Charge size can be controlled either volumetrically or gravimetrically. Gravimetric operation is preferred, as it offers superior control of charge size reproducibility and hence the degree of intensive mixing and melt temperature at discharge. The gravimetric feeding system must therefore be capable of delivering 8-12 kg of powdered material with a reproducibility of ±0.1% in less than 10 seconds while operating in a stop/start mode. Only recent advances in feeder technology have made this goal achievable.

PIGMENTS AND FILLERS

Another factor effecting quality is the surface area of the solid pigments and fillers and the quantity of resin "binder" available to wet out and encapsulate the material. The degree of dispersion is determined during the brief flux melting period when solid particles begin to embed themselves into and amalgamate with softened resin particles. Only minimal extensional shear is available to further modify the molten blend during the remaining few seconds of the cycle. As a first approximation, the volume fraction of solids in the formulation provides a crude guide to maximum loading levels achievable. In general, highly dispersed pigment formulations are possible at maximum volume fractions of 0.33 to 0.4 (calculated for 150°C). For example, LLDPE based N-762 carbon black formulations with excellent dispersion at 52 weight percent loadings exhibit rapidly deteriorating dispersion quality above about 54-55 weight percent (0.35-0.36 volume fraction). Titanium dioxide pigments can be successfully incorporated in LLDPE at levels up to 74-75 weight percent (0.36 volume fraction) before encountering poor dispersion (Figure 6).

Although surface areas and particle size distributions undoubtedly play a role in determining maximum loading levels, particle shape, density and heat capacity are also likely to play a role. The complex nature of the inter-relationships between these factors and the lack of suitable metrics has made it difficult to explore this phenomenon beyond a strictly empirical level, however. BET surface area estimates, oil absorption numbers and average particle

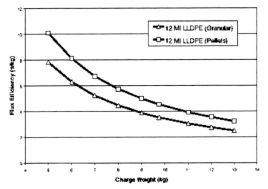

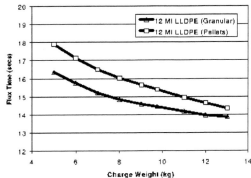

Figure 7. The effect of pelletized and granular LLDPE resin on HIDM flux times.

Figure 8. The effect of pelletized LLDPE resin vs. granular LLDPE resin on HIDM fluxing efficiencies.

size values have all failed to provide reliable correlation across a range of formulation components.

RESIN CHARACTERISTICS

The effect of the physical form of the resin component on HIDM flux rates is also significant. For example, Exxon LL6202.19, a pelletized, 0.924 density butene-based LLDPE with a melt index of 12 dg/min was compared with Exxon LL5202.09, an equivalent granular grade (<10 mesh) retrieved directly from the Unipol reactor. Both were processed in a 40 liter HIDM at various charge sizes. A flux efficiency factor, FE_x, was defined as:

$$FE_x = FT_x/W_x$$

where FT_x was the flux time (in seconds) and W_x is the charge weight (in pounds). Linear regression produced the following flux efficiency equations for granular (Fe_g) and pelletized (Fe_p) resins:

$$Fe_g = 59.6W^{1.2}$$

$$Fe_p = 69.7W^{1.2}$$

The ratio of flux efficiencies for granular and pelletized resins was 69.7/59.6 or approximately 1.17. This ratio is indicative of a roughly 17% increase in flux efficiency (i.e.,

Table 1. Color concentrate formulations for comparative compounding trials

Designation	Component	
MB-Yellow 1	LLDPE resin (12 MI)	50.0
	Titanium dioxide (PW 6)	25.0
	Organic yellow (PY 62)	16.0
	Calcium carbonate	8.0
	Zinc stearate	1.0
MB-Red 1	LLDPE resin (12 MI)	49.5
	Titanium dioxide (PW 6)	6.5
	Red iron oxide (PR 101)	14.0
	Organic red (PR 48:2)	20.0
	Calcium carbonate	10.0
MB-Blue 1	LLDPE resin (20 MI)	40.0
	Titanium dioxide (PW 6)	17.0
	Organic blue (PB 15:3)	9.0
	Calcium carbonate	33.0
	Zinc stearate	1.0
MB-Blue 2	LLDPE resin (12 MI)	44.0
	Titanium dioxide (PW 6)	17.0
	Organic blue (PB 15:3)	9.0
	Diatomaceous earth	20.0
	Chimassorb 944 FD HALS	9.0
	Zinc stearate	1.0
MB-Kraft 1	LLDPE resin (7.5 MI)	15.0
	LLDPE resin (12 MI)	15.0
	Titanium dioxide (PW 6)	37.0
	Yellow iron oxide (PY 42)	29.0
	Red iron oxide (PR 101)	3.5
	Carbon black (PBk 7)	0.5

throughput rate) associated with the use of granular LLDPE resin over pelletized LLDPE resin. The effect of LLDPE resin form on flux efficiency and flux time is illustrated in Figures 7 and 8.

In principle, increasing the charge size and reducing the flux time accordingly can compensate for the lower flux efficiency observed with pelletized resin. There are limits to this approach since the 40 liter HIDM becomes power limited at charge sizes of approximately 21 pounds for granular resin and 24 pounds for pelletized material. Extremely large charge sizes can also effectively flood the feed section of the mixer, increasing the average time required to transfer material into the mixing chamber. Finally, the capacity of downstream equipment must also be taken into account.

Trials with 50 and 70 weight percent titanium dioxide formulations have confirmed that flux efficiencies are reduced with pelletized resins feeds but charge size adjustments can compensate for differences. Dispersion quality is only slightly poorer with the pelletized feed, suggesting that at least with high density solids such as titanium dioxide, the surface area of the resin particle is not critical to achieving acceptable dispersion quality. Conceivably, softened resin pellets may become highly deformed and subdivided yielding surface areas comparable to the granular feedstock. The major drawback to pelletized feeds are increased time to flux, which may ultimately stem from the poor thermal conductivity of polyethylene and the need to generate and conduct heat at a diffusion limited rate.

EXPERIMENTAL

FORMULATIONS

Five commercial LLDPE and LLDPE/LDPE-based concentrate formulations were selected for comparison in HIDM and Twin Screw compounding trials (see Table 1).

Four of the formulations contained relatively high organic and inorganic pigment loadings in combinations known to result in sub-optimal dispersion levels. Formulations with and without stearate dispersion aids were represented. MB-Blue 2 was similar in composition to MB-Blue 1, but contained a 9% loading of Chimassorb 944 FD, a polymeric hindered amine light stabilizer from Ciba-Geigy known to interfere with pigment dispersion, resulting in loss of color strength.

In addition, the fifth selection (MB-Kraft 1) contained inorganic pigments at loadings of 70%. Since one of the pigment components (iron oxide yellow) undergoes a shift in color to a red shade if the melt temperature and shear rates are not properly controlled, the formulation acts as a crude probe of shear heating during processing.

MATERIALS

The LLDPE resins used were butene-based copolymers (0.924-0.926 density) made by gas phase polymerization and supplied off the reactor in granular form (approximately 10 Mesh). They contained minimal stabilization. The LDPE resin component was a 0.917 density ground extrusion coating grade material. The titanium dioxide pigment was a TMP coated chloride process rutile grade designed for general plastic applications. The calcium carbonate was a stearic acid treated grade with a mean particle size of 0.8 microns. All other pigments and additives were run as supplied.

All materials for the HIDM and twin screw trials were drawn from the same raw material suppliers, although different production lots were used.

PROCESSING

HIDM trials were conducted in Colortech's Brampton manufacturing facility, using a 40 liter processing line equipped with a downstream gear pump and Gala underwater pelletizing system. Pre-blends of pigments, fillers, additives and resins were prepared using a 1500 kg capacity ribbon blender. The blends were then metered volumetrically into the HIDM mixer. Rotor speed was fixed at 1750 RPM (42 m s^{-1} tip speed). Processing conditions and charge sizes were optimized for each product. The average HIDM mixing time was approximately 14 seconds and throughput rates were pump-limited.

Twin screw compounding work was carried out in Bingen, Germany on a Theysohn TSK 60 mm 36:1 L/D co-rotating twin screw extruder fitted with a Gala underwater pelletizer. A proprietary screw profile was employed. Barrel temperatures, screw speeds and throughput rates were optimized for individual formulations. All components were pre-blended in a high intensity Welex-style mixer for two minutes. The pre-blend was then volumetrically metered into the extruder.

A minimum of 500 kg of each formulation was produced.

SAMPLE ANALYSIS

All HIDM and twin screw color concentrate samples were letdown into blown film for evaluation of color, strength and dispersion quality. Standard 1.5 mil (37 μm) films were prepared by blending concentrate samples with 1.0 MI LLDPE resin at a letdown of 5.0 weight percent and extruding through a 37.5 mm laboratory film line at a blowup ratio of 2.4:1 and lay flat width of 20 cm. Standard operating procedures were observed throughout.

Chromatic characteristics were determined for the visible wavelength range (400-700 nm) using an X-Rite SP98 Portable Sphere Spectrophotometer (X-Rite, Inc) with D65 illumination, specular reflections included using a 10° standard observer. Film samples were evaluated as four layers against a standard white background. All CIE L*a*b*, color difference (DE$_{cmc}$) and apparent strength values were determined as averages of 5 readings using standard algorithms incorporated in X-Rite's QA-Master software.

Dispersion values were determined by counting the number of undispersed pigment particles in a 30x30 cm square of 1.5 mil (37 μm) blown film placed on a light box.

Relative pigment strength was determined by combining 2.0 grams of the pelletized concentrate with 2.0 grams of standard titanium dioxide pigment and 46.0 grams of 2.0 MI LLDPE resin and blending on two-roll mill for 3 minutes. Roll speed, spacing and temperature (150°C) were selected to allow adequate mixing with insufficient shear to affect pigment dispersion. The samples were then pressed into 3 mm plaques and measured using the sphere spectrophotometer.

The ash content, melt index and pellet density of the HIDM and twin screw samples were also evaluated using standard ASTM methods.

RESULTS AND DISCUSSION

CIE L*a*b* color values for 5.0% blown film samples are shown in Table 2. Relative color strength and dispersion results are shown in Tables 3 and 4, respectively. Tint strength test results appear in Tables 5 and 6.

L*a*b* values in film and tint plaques were consistent across both compounding platforms, with overall color differences (DE$_{cmc}$) at or below 0.5 levels. No significant color shifts were noted in the samples tested, including the heat and shear sensitive MB-Kraft 1.

Table 3. Relative color strength in blown film samples

	% Color strength (apparent)	
	HIDM	Twin screw
MB-yellow 1	100.00	97.37
MB-red 1	100.00	103.00
MB-blue 1	100.00	96.14
MB-blue 2	100.00	98.18
MB-kraft 1	100.00	100.40

Table 4. Dispersion analysis of blown film samples

	Dispersion (specks/30x30 cm)	
	HIDM	Twin screw
MB-yellow 1	40 (gritty)	3
MB-red 1	11	15
MB-blue 1	21	29
MB-blue 2	17	22
MB-kraft 1	8	13

Table 2. Spectrophotometric color analysis results of blown film samples

	HIDM			Twin screw			DE_{cmc}
	L*	a*	b*	L*	a*	b*	
MB-yellow 1	87.70	2.11	71.48	88.37	1.44	72.05	0.50
MB-red 1	42.63	39.83	18.86	42.24	39.54	19.14	0.35
MB-blue 1	50.62	-16.35	-41.97	51.27	-16.68	-41.97	0.35
MB-blue 2	49.82	-16.19	-40.86	50.34	-16.40	-41.56	0.39
MB-kraft 1	63.09	12.57	28.08	63.08	12.90	31.16	0.28

Table 5. Concentrate tint strength

	HIDM			Twin screw			DE_{cmc}
	L*	a*	b*	L*	a*	b*	
MB-yellow 1	90.36	-0.92	53.63	91.06	-1.01	53.78	0.26
MB-red 1	53.14	38.50	5.28	52.94	38.42	5.54	0.20
MB-blue 1	62.07	-22.21	-37.48	62.63	-22.55	-43.26	0.43
MB-blue 2	63.07	-21.17	-34.27	62.89	-21.52	-34.84	0.30
MB-kraft 1	70.99	8.53	22.21	71.56	8.4	22.00	0.26

In four of the five formulations tested, HIDM samples showed marginally better dispersion levels in blown film than the twin screw materials. The single exception was MB-Yellow 1, where a dispersion rating of 40 was measured for the HIDM product. In contrast, the twin screw version of MB-Yellow 1 provided the best film dispersion quality noted during the trial.

Table 7. Comparison of miscellaneous concentrate properties

	Ash content, %		Pellet density, g/cm³			Melt index g/10 min, 190°C, 2.16 kg	
	HIDM	Tween screw	Calculated	HIDM	Twin screw	HIDM	Twin screw
MB-yellow 1	38.0	38.3	1.35	1.23	1.37	6.2	9.4
MB-red 1	33.1	34.9	1.26	1.17	1.28	18.1	22.2
MB-blue1	53.7	52.2	1.50	1.37	1.47	8.1	12.2
MB-blue 2	40.1	41.8	1.31	1.16	1.30	7.8	7.7
MB-kraft 1	66.6	67.2	2.01	1.88	2.00	<0.1 (34.9)*	<0.1 (45.0)*

*190°C, 10 kg

Table 6. Relative color strength in tint samples

	% Color strength (apparent)	
	HIDM	Twin screw
MB-yellow 1	100.00	95.71
MB-red 1	100.00	101.55
MB-blue 1	100.00	96.47
MB-blue 2	100.00	104.02
MB-kraft 1	100.00	95.46

This may be a function of the higher levels of shear stress available to the twin screw operator though control of fill volume and throughput rates. Interestingly, superior pigment dispersion did not translate into significant increases in relative color strength values in film or in tint tests for MB-Yellow 1.

In general, the relative apparent color strength values calculated for all of the twin screw samples fell between 95.46 and 104.02% of the corresponding HIDM samples. A variation of ±5% is considered within experimental error and is generally not considered significant. This is based on the observation that many pigment manufacturers consider their products to be within specification when color strength is within ±5 to 10% of target. Color strength trends in film and tint plaques are consistent between HIDM and twin screw samples, with the exception of MB-Blue 2 which shows somewhat higher values in tint and MB-Kraft 1 which trends slightly lower in tint.

Ash testing (shown in Table 7) measures non-combustible residues left behind when concentrate samples are burned in air under controlled conditions. They are indicative of the original composition of the sample and can be used to evaluate the degree to which components have been successfully incorporated or lost to dust collection systems, etc. No clear trends emerged from these trials with HIDM and twin screw samples exhibiting similar variability, however.

Pellet density values are a measure of the degree to which entrapped air and/or volatiles have been successfully removed from the formulation during compounding and pelletizing. Entrapped air results in voids that may or may not be visible to the naked eye when pellets are cut open. In this study, twin screw samples yielded density values much closer to the theoretical values calculated for the formulations. This was presumably due to the vacuum venting applied to downstream barrel segments during the compounding operation. This work led to several refinements of the HIDM gear pump system to improve de-gassing and reduce the void inclusion level in subsequent production.

CONCLUSIONS

The results of this study clearly indicate that both HIDM and twin screw compounding technologies can provide concentrate products with virtually identical quality when operating conditions are properly optimized. Where differences exist, they appear to be formulation specific. Both technologies offer equivalent ability to disperse organic and inorganic pigments effectively, although the twin screw extruders provides the capability to increase shear stress if required.

ACKNOWLEDGMENTS

We wish to acknowledge and thank Wil Jonkers and the manufacturing team at PPM (Bingen) for their valuable contributions to this study. We would also like to thank Felix Calidonio, Rose Wainwright and Rob Peterson who performed the bulk of the analytical work.

REFERENCES

1.   E. L. Canedo, N. L. Valsamis, *Intern. Polymer Processing IX*, **3**, 225 (1994).
2.   **U.S. Patent No. 4501543**, G. J. Hahn, R.N. Rutledge.
3.   Z. Tadmor, P. Hold, L. Valsamis, 37th SPE ANTEC Technical Papers, 193 (1979).
4.   P. Hold, Z. Tadmor, L. Valsamis, 37th SPE ANTEC Technical Papers, 205 (1979).
5.   Draiswerke GmbH (a member of the Eirich Group), P.O. 31 02 20, D-68262 Mannheim, Germany.
6.   **U.S. Patent No. 4420449**, A. N. Wright and others.
7.   R. Adams, Z. Crocker, A.N. Wright, 40th SPE ANTEC Technical Papers, 640 (1982).
8.   K. C. Chu, R. T. Woodhams, A. N. Wright, 43th SPE ANTEC Technical Papers, 699 (1985).
9.   W. Baker, P. Patel, A. Catani, 44th SPE ANTEC Technical Papers, 1205 (1986).
10.   D. Lyons, W. E. Baker, *Intern. Polymer Processing V*, **2**, 136 (1990).
11.   S. Frenken, D. Lyons, W. E. Baker, 49th SPE ANTEC Technical Papers, 40 (1991).
12.   **U.S. Patent No. 4820469**, M. J. Walsh, E. G. Maury and others.

In-line Color Monitoring of Pigmented Polyolefins During Extrusion. I. Assessment

Ramin Reshadat and Stephen T. Balke
Department of Chemical Engineering and Applied Chemistry, University of Toronto, ON
M5S 1A4, Canada
Felix Calidonio and Christopher J. B. Dobbin
Colortech Inc., Brampton, ON L6T 3V1, Canada

ABSTRACT

The color of pigmented polyolefin melts was measured in-line during processing by using a fiber optic equipped visible-near-infrared spectrophotometer. Pigment loadings varied from 3.6 to 56 wt.% with some formulations containing up to 33.1 wt% calcium carbonate filler. Results showed that within specification color could be distinguished from off specification color. Also, pigment degradation and temperature tolerance could be assessed. Predicting needed off-line measured color from these in-line results will be the subject of Part II of this series.

INTRODUCTION

In the manufacture of color master batches, the current industry practice is to perform critical color measurements off-line. Typically, a sample of the pelletized color concentrate is diluted with natural resin at a standard ratio and milled, extruded or injection molded into a physical form suitable for visual and instrumental evaluation. These methods are slow and labor intensive. Furthermore, they do not lend themselves well to statistical process control strategies because of the time lag between production and testing. Since relatively few samples can be examined, laboratory measurements may not give a true indication of the consistency of the concentrate product over the entire manufacturing process.

Most "in-line" color measurement techniques currently in use are more correctly referred to as "on-line" methods. Usually, they consist of some form of sampling device that conveys pelletized masterbatch from the production line to an optical cell. There, an auto-mated spectrophotometric system attempts to read a reflectance curve from the surface of the pellets. Changes in pellet size and surface texture can have an adverse effect on the batch-to-batch reproducibility of the color reading.

The in-line process monitoring technique discussed here involves color measurements performed directly on the pigmented polymer melt, ideally while still in the compounding ex-truder. In contrast to the usual post-production methods, the in-line technique can offer almost instantaneous analysis results. In addition, the entire production run can be sampled continu-ously.

We have previously shown that a fiber optic equipped near-infrared spectrophotometer (NIR) can be used to measure the composition of a blend of polypropylene with recycled polyethylene.[1] In this current work, we show the results of monitoring color in-line, rather than composition, by using the visible part of the spectrum measured from the same spectrophotometer (which we now term a Vis-NIR instrument).

Our primary objective is to determine whether slight differences in the color of heavily pigmented polymer melts can be successfully differentiated by the Vis-NIR system. We will also briefly examine thermochromic phenomena and the effect of melt temperature on color. Finally, the problem of relating color differences measured in-line to data obtained off-line from the same masterbatch when fully diluted in an end-use resin at room temperature will be discussed.

THEORY

The most commonly used laboratory instrument for the precise measurement of the reflected light (color) is the spectrophotometer. Typically, this instrument is used to record a visible wavelength (400-700 nm) reflectance spectrum which acts as a fingerprint of the colored sample being characterized. The shape of the spectral curve is influenced by the number and types of colorants employed in the sample, the nature of the illuminating light source and the spectral sensitivity of the detector (observer) system.

Once a reflectance curve is obtained spectrophotometrically, the measured information is mathematically transformed according to standard conventions, into the numbers used to describe the color of the sample. In practice, a standard observer and standard illuminant are selected based on test methods defined by the Commission International de l'Eclairage (CIE). Calculation procedures involve the numerical integration of the product of the spectral power distribution $S(l)$ of the light source and the reflectance factor $R(l)$. Reflectance factors repre-sent the percentage of light reflected by the sample at each wavelength.

Weighting factors published in the American Standard Test Method, ASTM E 308-85, were used to calculate tristimulus values from spectral reflectance data. The weighting factors selected were those corresponding to CIE 10° Standard Observer and Standard Illuminant A. Since Standard Illuminant A exhibits a spectral power distribution curve with a color temperature of 2856°K, the output of a Halogen-Tungsten lamp could be used to approximate this light source.

Once tristimulus values have been obtained for each reflectance curve, they may then be used to calculate a unique set of three co-ordinates representing the sample color in CIE L*a*b* color space. The conversion equations are straightforward and may be performed on a personal computer.[2] In this three-dimensional space, the L* co-ordinate represents the lightness/darkness of the sample and has values between 0 for black and 100 for white samples. The coordinates a* and b* represent the position relative to the red/green and yellow/blue axis respectively with values falling between +60 and -60. Figure 1, shows a typical representation of the CIE L*a*b* color space.

Based on Figure 1, the more positive the a* co-ordinate, the more red the sample. The more negative the b* co-ordinate, the more blue the sample, etc. A simplified 2-dimensional plot of b* versus a* will be used in this study to illustrate color differences between samples.

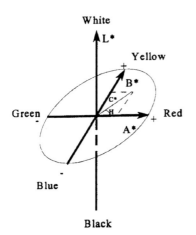

Figure 1. CIE L* a* b* color space.

EXPERIMENTAL

EXTRUSION AND MONITORING

A counter-rotating, fully intermeshing twin-screw extruder (model 2008 C. W. Brabender Instruments Inc.) was used to extrude a series of commercially prepared masterbatch samples. The extruder was equipped with three heating zones; two located in the barrel and one in the die.

A model 6500 On-Line Vis-NIR spectrophotometer (NIR Systems, Inc., Silver Spring, MD) equipped with single bundle bifurcated fiber optics was used for data acquisition. An Interactance Reflectance probe was used along with a novel extruder/monitor interface design to obtain diffuse reflectance spectra of the plastic melt near the die. The visible region of the spectrum (400-700 nm) was used for analysis. Figure 2 shows a schematic of the experimental setup.

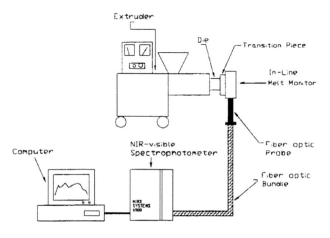

Figure 2. Experimental equipment setup.

MATERIALS

A representative selection of commercial color concentrates (Colortech Inc., Brampton, ON) were used for this study. Each pelletized sample consisted of a blend of organic and inorganic pigments dispersed in Linear Low Density Polyethylene (LLDPE) carrier resin. Pigment loading varied between 3.6 and 50.0 wt.% and some formulations contained up to 33.1 wt.% calcium carbonate filler. All were prepared on a production-scale 40 Liter Gelimat-style high intensity dispersive compounding line.[3,4]

In addition to so-called good concentrate samples exhibiting "on-spec" color, several formulations were also supplied in an "off-spec" form. These materials consisted of nominally identical formulations that had suffered minor weigh-up errors, contamination or insufficient pigment development during manufacture. They were thought to represent a realistic test for color discrimination trials.

All materials were extruded "as-is" without dilution. Masterbatch formulations were coded as follows:

"MB- Apparent color Apparent lightness Quality"

Apparent lightness/darkness is indicated by either 1 (light) or 2 (dark). Color quality is shown as G (good, within specification) or B (bad, off-specification). For example, MB-Green 1G refers to a light green formulation exhibiting acceptable color while Gray 2B designates a dark gray with poor, off-spec color.

EXPERIMENTAL METHOD

In order to compare the Vis-NIR instrument equipped with a diffuse reflectance probe to a standard laboratory color instrument (in this case, an X-Rite SP-68 integrating sphere spectrophotometer), off-line measurements were carried out on pressed plaques at room temperature. All color values were calculated using a standard A illuminant and a 10° observer angle.

In-line measurements of the polymer melt were obtained in the temperature range of 150 to 200°C. The extruder speed was set to 30 rpm throughout. Samples were allowed to run through the system for 15-20 minutes prior to spectral collection. Ten spectra were obtained for each sample over a period of 10 minutes. Each spectrum required about 18 seconds to be obtained and is the average of 35 individual sample scans.

RESULTS AND DISCUSSION

OFF-LINE RESULTS

Table 1 shows the L*, a* and b* values (Standard illuminant A/10° standard observer) for ten samples analyzed with the two instruments off-line. Overall, it is immediately evident that both instruments can distinguish good colors from bad. However, results from the two instruments are not identical. On the L* plane, the fiber optic equipped Vis-NIR instrument yields a consistently higher lightness than the SP-68 benchtop spectrophotometer. Since the sapphire window at the end of the probe, transmits a maximum of only 85% of the incident light at 550 nm, a maximum of 15% of the light will be reflected back to the detector as specular reflection. This is accompanied by the light reflected back from the melt. Thus the total reflectance is increased, resulting in a higher L*. At wavelengths lower than 550 nm and wavelengths higher than 770 nm, the transmission loss in the sapphire is higher, so the reflection at these wavelengths is more. Thus, a* values (an indication of the redness of the sample) are lower (except for the gray and green samples that reflect the light in almost all regions of the spectrum), and b* (an indication of the blueness of the sample) are significantly lower than those from the standard instrument.

This same systematic bias is expected to apply to the in-line measurements described in the next section.

Table 1. Off-line color measurements. Concentrate plaque samples

Sample	X-Rite (A/10°)			Vis-NIR (A/10°)		
	L*	a*	b*	L*	a*	b*
MB-green 1G	69.39	-36.64	-10.63	76.39	-36.46	-18.11
MB-green 1B	68.99	-33.55	-5.53	76.49	-33.43	-14.26
MB-gray 1G	55.41	-2.59	-3.77	60.81	-3.06	-11.35
MB-gray 1B	54.60	-3.77	-3.54	59.71	-3.13	-10.86
MB-gray 2G	43.64	1.27	-0.32	46.95	0.16	-5.70
MB-gray 2B	41.86	-1.24	-1.66	43.89	-2.85	-8.02
MB-red 1G	41.45	28.25	23.58	46.78	21.41	16.80
MB-blue 1G	38.06	-16.98	-44.94	43.37	-23.83	-53.13
MB-green 2G	36.18	-18.63	2.19	37.65	-18.88	0.33
MB-orange 1G	57.39	40.21	52.40	65.32	32.44	45.11

IN-LINE RESULTS

Table 2. Vis-NIR in-line color measurements (melt temperature = 150°C)

Sample	L*	a*	b*
MB-green 1G	55.87±0.44	-12.56±0.07	-10.36±0.07
MB-green 1B	55.21±0.03	-12.24±0.05	-8.53±0.09
MB-gray 1G	50.65±0.05	-3.39±0.10	-8.97±0.12
MB-gray 1B	50.48±0.02	-3.03±0.08	-7.51±0.14
MB-gray 2G	43.03±0.06	-0.99±0.15	-5.16±0.09
MB-gray 2B	42.42±0.01	-1.92±0.14	-6.08±0.13

Table 2 shows the L*, a* and b* values and their 95% confidence interval calculated at 150°C for three pairs of extruded masterbatch samples. Each pair contained one "good" and one "bad" concentrate formulation. Although the reflectance spectra obtained from good and off colors almost overlap each other, the color values generated from the reflectance data can be success-fully differentiated. That is one of the advantages of working in CIE L*a*b* color space. Al-though L*, a* and b* differences between good and bad MB-Green 1 and MB-Gray 2 pairs derived from the Vis-NIR agrees well with SP-68 reference data, more subtle color difference such as that exhibited by the MB-Gray 1 pair may tax the capabilities of the current on-line system.

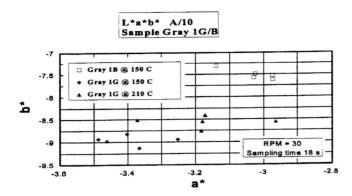

Figure 3. b* vs. a* plot for sample MB-gray 1G/B, at different temperatures.

The effect of extrusion melt temperature in b* and a* values on sample MB-Gray 1G is illustrated in Figure 3. This formulation consists of 48 wt% rutile titanium dioxide (TiO$_2$), with small quantities of carbon black and yellow iron oxide (1.0 and 1.2 wt.% respectively) dispersed in a LLDPE resin carrier.

It was observed that both a* and b* values increased as the melt temperature at the die was incrementally raised from 150 to 210°C. As indicated, a* was seen to shift towards the red region while b* moved into the yellow. It is tempting to interpret this color shift as pig-ment or polymer degradation. The yellow iron oxide component could conceivably have lost water of hydration to yield the thermally stable red iron oxide form. Experience suggests, however, that this pigment is stable to temperatures of approximately 225-230°C under similar extrusion conditions (i.e., extruder residence times of less than 2 minutes). In fact, subsequent examination of the extrudate from this run indicated that MB-Gray 1G reverted back to its original color when cooled to room temperature. Temperature induced color shifts of this magnitude were not observed in formulations in which TiO_2 was not a major component.

We attribute the temperature-induced color shift in MB-Gray 1G to a reversible thermochromic phenomenon taking place in the titanium dioxide pigment component. Transient yellowing of hot TiO_2 concentrate formulations is a commonly observed event during the manufacture of masterbatches but the theoretical aspects of these color shifts are poorly understood. Anecdotal evidence suggests that similar reversible shifts occur with some organic yellow pigments such as diarylide yellow (Pigment Yellow 13), although solubility and recrystallization effects may be involved here.

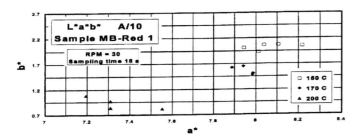

Figure 4. Effect of increasing temperature on the color of sample MB-red 1.

Figure 4 illustrates the effect of increasing melt temperature on the color of MB-Red 1G. This formulation consists of red iron oxide (Pigment Red 101) at 20 wt%, calcium 2B red (Pigment Red 48:2) at 10 wt% and titanium dioxide at 5 wt% in an LLDPE carrier. Replicate measurements clearly show a shift to lower a* and b* values corresponding to the development of a slightly bluer/greener cast at 200°C.

Subsequent lab tests were conducted by extruding MB-Red 1G onto a polyester sheet and monitoring the L*a*b* characteristics as the melt cooled using the X-Rite spectrophotometer. The melt exhibited a reverse color shift towards higher a* and b* values as the temperature dropped from 200 to 100°C. Confirming that this was also a reversible thermochromic change. Finally, oven heating trials with raw pigment samples identified the thermochromic agent as the calcium 2B red component. Heating this pigment to 200°C resulted in a dark brick red that once again reverted to its normal bright red color on cooling to room temperature.

One of the more intriguing aspects of in-line color monitoring should be the ability to monitor and control color in the final product. By comparing Table 2 with Table 1 it can be seen that there are evident differences between the color numbers in the off-line and in-line measurements as well as between instruments. As mentioned above, color perception greatly depends on the sample. Shape, physical state and temperature of the sample will alter the intensity of the reflected light from a sample and hence its color. Thus, it is not surprising that the measured color of the melt and that of the finished product are not equal. The problem to be solved is how to draw a correlation between the two. Methods of accomplishing this include partial least squares as well as neural network modelling.

CONCLUSIONS

It is feasible to use a fiber-optic-assisted visible spectrophotometer to measure the color of a stream of molten pigmented polymer in-line. In-line melt monitoring can distinguish within specification color from out of specification color. Also, it can be used to detect pigment degradation and to determine the upper temperature thresholds for pigmented polymer processing. However, because of the many factors that determine color perception, off-line and in-line results are not equivalent.

ACKNOWLEDGMENT

We wish to thank NIRSystems, Inc. (Silver Spring, MD.) for supplying the fiber optic equipped Vis-NIR instrument. Also, we are grateful to Colortech Inc. (Brampton, ON) for providing pigmented plastics and off-line results. Finally, we wish to thank the Ontario Centre for Materials Research for funding this project.

REFERENCES

1   R. Reshadat. W. R. Cluett, S. T. Balke, J. W. Hall (1995), "Quality Monitoring of Recycled Plastic Waste during Extrusion: I. In-Line Near-Infrared Spectroscopy", ANTEC'95 Proceedings.
2   A.Beger-Schunn. (1994), **Practical Color Measurement**, *John Wiley and Sons*.
3   R. Adams. Z. Crocker, A. N. Wright (1980), SPE 40th ANTEC, p. 640.
4   W Baker, P. Patel, A. Catini (1986), SPE 44th ANTEC, p.1205.

The Effects of Injection Molding Parameters on Color and Gloss

Eric Dawkins, Kris Horton, Paul Engelmann
Department of Industrial and Manufacturing Engineering, Western Michigan University,
Kalamazoo, MI, USA
M. Monfore
Ralston Foods

ABSTRACT

Within the plastics industry there is a growing trend towards producing unpainted finished products. This places an increased emphasis on understanding which process variables can induce shifts in color. This study focused on ABS parts using a newly specified automotive red colorant. A 1/8 fractional factorial design of experiment was used to test the effects of seven independent variables upon color and gloss. The results showed which parameters were responsible for influencing color and gloss. Also determined was the color range achievable within the process window.

INTRODUCTION

The automotive industry is an extremely competitive market for the injection molding industry. Automotive manufacturers are reducing the number of suppliers which produce components for their vehicles. To stay competitive, injection molders need to be proactive in both technology and pricing. Interior trim is a very competitive market for the injection molding industry. Much of the interior of an automobile is comprised of plastic parts molded from many different materials. The variety of plastics used in interior trim makes it very difficult for the assembler to obtain color match between all components. Each type of base resin requires a different formulation of colorant to achieve the same finished color. This problem is further complicated by the addition of additives such as UV stabilizers, lubricants, fillers, etc. Other problems include molding faults such as flow lines, weld lines, sink, splay, and blush which are highly visible on show surfaces. Therefore, much of the interior is painted to achieve a consistent color and gloss across the surface of the interior. Although automotive manufacturers recognize these problems exist, molders are asked to eliminate them so that expensive painting operations can be avoided.

A supplier of automotive trim components was in the process of engineering a new product along with a new colorant. The colorant manufacturer supplied several iterations of colorant to the molder with little success at achieving the customer's requirements. This raised the question "How can we influence color and gloss through the injection molding process?"

There are numerous variables within an injection molding system which might influence color or gloss. The influence may be perceived in the form of color shift or color variation. The challenge lies in determining how and to what degree each variable or combination of variables contributes to this effect. If the correlation between process parameters and color is significant, injection molders could gain more understanding in molding products with difficult color and appearance specifications. Therefore, the hypothesis tested by this experiment was to determine which independent variables have the most significant effect on color and gloss.

METHODOLOGY

MATERIALS

The molding material used was a natural DOW Magnum 344 resin. Colorant was a Toreador Red supplied by PMS Consolidated. Red colorant was chosen in part because it is widely used in industry and is notoriously difficult to color match.[1] The colorant was blended with the resin at a 25:1 ratio and was batch blended to assure uniform consistency. Drying was performed according to the manufacturer's recommendation of 82°C and a moisture content of 0.04%.

EQUIPMENT

Molding was performed on a 1992 77 metric ton (85 ton) Van Dorn injection molder with an EL controller. A non-vented barrel and a 35 mm (1 3/8") general purpose screw with a length to diameter (L/D) ratio of 20:1 was used. The intensification ratio for the machine was 10:1. A 1993 AEC mold temperature controller and a 1995 2.7 metric ton (3 ton) Thermal Care Chiller were used to control water temperature in the mold. The mold used was a standard ASTM tensile bar mold. Resin was dried using a 1990 Una-Dyn UDC style dryer with an OMNI II-X controller and a digital dew point meter. Material loading was done using a 1993 AEC hopper loader. Process monitoring and data acquisition was performed using a RJG Technologies DART system. Cavity pressure, screw position, and hydraulic pressure were monitored.

EXPERIMENTAL DESIGN

Independent variables thought to have a significant effect on color and gloss were selected based upon past research and molding experience.[2,3] A 1/8th fractional factorial experimental plan was used to test the effects of seven independent variables upon the five dependent variables listed below. The experimental plan was developed and analyzed using Statistica® software.[4]

Independent Variables	*Dependent Variables*
coolant temperature	L*
back pressure	a*
cycle time	b*
screw speed (rotation)	gloss
injection speed	part weight
barrel temperature	
pack/hold pressure	

Barrel temperature profiles were used as the set points in the process since it was not possible to set the melt point. The resulting melt temperatures were recorded at each setting. Coolant temperature settings were used to control mold temperature. Pack and hold pressures were paired and represented as a single independent variable. Resulting melt pressure at the nozzle was estimated using the intensification ratio of the press.

PROCEDURE

Process parameters were established by setting the press to achieve the manufacturer's recommended processing conditions for the Dow 344 material. These variables included melt temperature, pack pressure, back pressure, and mold temperature. On-machine rheology curves were produced to establish injection fill speed ranges. A gate seal study was performed to establish minimum pack and hold time requirements. High and low parameter settings were then established by exploring the range for each variable until acceptable parts could be produced at all settings. The variables were then inserted into the experimental design and the runs arranged randomly. A total of 19 runs were performed. Two of the runs were replicated to allow estimation of experimental error. Part weight was used to determine process stability before samples were saved. Part weight was measured at press side and logged into a computer spreadsheet for trend analysis. The experiment was performed over a consecutive 12 hour period.

COLOR MEASUREMENT

Color was measured using a Macbeth 1500/Plus Spectrophotometer with an Illuminate D, observer of two degrees area view and Quick Key® software. Quick Key uses the CIELAB

Color Space system (ASTM 308 Standard) which is widely used in the industry. It provides 3 color readings that occupy the CIELAB color space: L*, a*, and b* values.[5]

L* provides a measure of light/dark.

a* measures the degree of red/green.

b* measures the degree of yellow/blue.

Sample preparation and measurement procedures for spectrophotometric analysis must remain consistent in methodology to produce accurate data.[6,7] Therefore, a fixture was designed to hold the tensile bars on the spectrophotometer. Color measurements were performed after parts were allowed to stabilize for three days in controlled conditions following the experiment. The color measurements (L*, a* and b*) were taken on the cavity side of the part opposite of the ejector pins. Measurement locations included the left, center and right locations on the surface of the part. Gloss was measured using a microprocessor enhanced version of the Statistical Novogloss meter. Its capabilities include auto-ranging from 0.1 to 1000 gloss units at 60°. Gloss readings were taken on the left and right surface locations of the part. The surface area near the center of the part was too small to obtain an accurate reading with the gloss meter. Data for L*, a*, b* and gloss values were entered into a spreadsheet and imported into Statistica® software for data analysis. Part weight was also recorded to correspond with color measurements.

FINDINGS

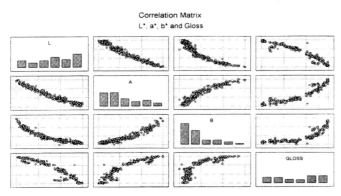

Figure 1. Correlation matrix. L*, a*, b*, and gloss.

A matrix of scatterplot graphs were produced to explore the correlation between L*, a*, b*, and gloss. This is shown in Figure 1. It can be seen that positive correlation exists between a*, b*, and gloss values. L* is negatively correlated with a*, b*, and gloss. This means, that as L* increases, the values of a*, b*, and gloss decrease. Conversely, if either a*, b*, or gloss increases, the remaining two will increase as well. At the same time the value for L* will decrease.

Data was then analyzed using standard analysis of variance methods. Substantial changes in color and gloss due to process variables were evident. The changes in color and gloss were easily observed with the naked eye. The analysis showed that the maximum color

Table 1. P values of effects and responses

Dependent variables	Independent variables						
	Coolant temperature	Barrel temperature	Injection speed	Pack pressure	Screw speed	Cycle time	Back pressure
L*	***	***	***	**	---	---	---
a*	***	***	*	*	---	---	---
b*	***	***	---	---	---	---	---
Gloss	***	***	***	**	*	---	---

*** P=0.001, ** P=0.01, * P=0.05, --- not significant

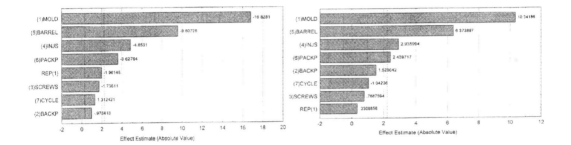

Figure 2. Pareto chart of standardized effects; variable L*; 7 factors at two levels; MS residual = 0.2265886; p = 0.05.

Figure 3. Pareto chart of standardized effects; variable a*; 7 factors at two levels; MS residual = 0.1368829; p = 0.05.

shift (response range) throughout the experiment was 9.02 for L*, 4.77 for a*, 3.24 for b*, and 57.02 for gloss.

Results of independent variables and their effect on the dependent variables are summarized in Table 1. p-values are shown for the independent variables that are significant. Results were further broken down into Pareto charts of standardized effects for each dependent variable as shown in Figures 2-5. Independent variables were considered significant at p <= .05.

Pareto effects for L* are shown in Figure 2. It was shown that mold temperature (MOLD) had the largest effect on L*. Mold temperature had almost twice the effect of barrel temperature (BARREL). These were followed by injection speed (INJS) and pack pressure (PACKP).

Figure 3 shows that a* was effected most by mold temperature followed by barrel temperature. Injection speed and pack pressure had the least effect of the significant variables.

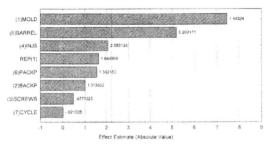
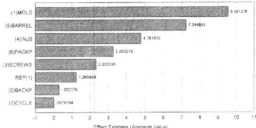

Figure 4. Pareto chart of standardized effects; variable b*; 7 factors at two levels; MS residual = 0.1182665; p = 0.05.

Figure 5. Pareto chart of standardized effects; variable gloss; 7 factors at two levels; MS residual = 18.74314; p = 0.05.

Mold temperature was shown in Figure 4 to have the largest effect on b*. Barrel temperature was the only other variable to show significance.

The Pareto chart for gloss is shown in Figure 5. The largest effect was mold temperature. Barrel temperature was next followed by injection speed, pack pressure and screw speed (SCREWS).

CONCLUSIONS

Coolant temperature (mold temperature) was shown to have the greatest effect upon color and gloss followed by barrel temperature (melt temperature), injection speed, and pack pressure. The ranking of effects for the process variables was the same for all color responses. This is very convenient, but at the same time, could create undesired problems when trying to adjust the process to achieve a certain response value. This is illustrated in Figure 1. If change is created in either L*, a*, b*, or gloss, the remaining constituents will likely change as well. Therefore, if only one of the color variables is not within specifications, it may be difficult to achieve the desired value as the remaining variables may shift as well. However, if the color or gloss values are off as a whole, it may be feasible to achieve the desired results by manipulating process variables. This will require follow-up experiments to confirm these effects.

Generally accepted tolerances for color difference from a master plaque are 0 +/- 1.0 for L* and 0 +/- .5 for a* and b*.[8] The range of responses listed above for L*, a*, and b* are much greater than the allowable tolerance.

Further experimentation is needed to understand the interaction between process variables and responses. The effects of resin, additives and colorants, and mold geometry also need to be explored in a follow-up experiment.

The main effects for a product that are being molded may or may not be the same as these. The results of this experiment reveal that color and gloss are affected by process variables. Future experimentation will be necessary to investigate effects with other materials and mold geometry.

ACKNOWLEDGMENTS

We would like to thank members of the 1995 Advanced Plastics Processing class at Western Michigan University: Raj Krishnan, Roger Krontz, Frank Rinderspacher, Rick Truza, Ryan Wejrowski and Chong Wong. Each member made significant contributions in the research and experimentation stages of this project. Tony Kiszka was very helpful in teaching us how to operate the spectrophotometer. Special thanks is also extended to Summit Polymers, Inc. who contributed colorant and the use of color analysis equipment.

REFERENCES

1. C. Brown, M. Van der Kooi, S. Ramrattan, The effects of vented and non-vented plastication systems on color and specific gravity. Technical Papers, vol. 41, pp. 586-589. Brookfield, CT: SPE (1995).
2. F. Furches, J. Bozzelli. Screw design efficiency in color distributive mixing with abs resins. Technical Papers, vol. 33, pp. 6-9. Brookfield, CT: SPE (1987).
3. H. Sirett, J. Suthers,. Polyolefin master plaque approval for the automotive industry. Elements of Color & Appearance for the Thermoplastics Processor, (RETEC Proceedings) pp. 15-34. Brookfield, CT: SPE (1991).
4. StatSoft, Inc., STATISTICA for Windows Tulsa, OK 74104 (1996).
5. X-Rite, A Guide to Understanding Color Communication. Grandville, MI(1990).
6. Macbeth. Series 1500 Color Measurement System: Operator's Manual. Newburgh, NY, (1982).
7. Hunter Lab. Instruments for Color and Appearance Measurement. Reston, VA (1988).
8. K. Shellnut, Chrysler's color control program. Elements of Color & Appearance for the Thermoplastics Processor, pp. 1-14. Brookfield, CT: SPE, (1991).

Method for Effective Color Change in Extrusion Blow Molding Accumulator Heads

J. S. Hsu, D. Reber, Cincinnati Milacron, 4165 Halfacre Road, Batavia, OH 45103, USA

ABSTRACT

The design of accumulator blow molding heads to promote quick and effective color and material change is examined. Quick color and material change is essential for processors, particularly custom blow molders, to reduce the non-productive machine time and scrap materials generated. The design approach uses basic flow characteristics dealing with temperature-viscosity relationships of the flow channel profiles and conditions of the flow surfaces in the head. Test and photographic data are used to support the findings.

BACKGROUND

The problem of color change in the "large part blow-mold process" often relates to "streaking" on parisons. These streaks tend to fade away slowly. Accumulator heads with better color change features may take a few hours. Streaking could sometimes last over 24 hours for some less streamlined designs, to change from dark to the opposite end of the color spectrum (e.g., black to white). The larger the head, the longer it takes.

For custom molders who change color on a regular basis, this becomes a costly issue. Besides losing valuable productive machine time and labor, it can also generate from a few hundred to a few thousand pounds of scrap which needs to be handled, ground-up and re-processed. In today's highly competitive market-place, fast color change for large part industrial blow molders is the extra edge one cannot afford to overlook.

SOURCES OF PROBLEM

Before improvements can be made, sources of the streaking problem have to be identified and located. Streaking initiates from somewhere along the melt flow-channels inside die heads and/or manifolds. They are the "bleed-off" from pockets of residual or stagnant material located behind the slow moving areas or corners.

Pockets may also be found around the melt distribution (spreading) sections. This is due to unbalanced flow when processing near the limits or outside the "designed operating range" for a long period of time (e.g., on large industrial blow mold machines, using less than ~1/3 of shot capacity, molding heavy-wall small parts, with long cycle time, etc.)

ANALYSIS

Flow channel sections, melt and flow characteristics of a given material, and melt and flow characteristics between materials, all have an influence on how well a color/material change is performed.

FLOW CHANNEL SECTIONS

In an accumulator head, there are many different shapes, sizes and configurations of flow-channel sections connected to one another, either in series or in parallel. At the same time, melt is being split, shaped, turned, directed, bent, distributed, re-knitted, and finally, is collected inside a large diameter. Pressure is the only prime mover that forces the stream of melt through all these passages.

As it flows from section to section, changes in channel shape and size causes melt velocity to speed up or slow down. When velocity increases, flow is directed and guided; when velocity decreases, flow wanders and loses direction. Typically at bends or turning corners, pockets of stagnant material fills the voids outside the main melt stream.

MELT AND FLOW CHARACTERISTICS

Melt viscosity dominates the way the melt flows, and yet it can be influenced by many other factors and flow conditions. Resin and grade (density, MW, MWD, M.I., etc.) are of course the main factors that determine melt viscosity. Colors and additives, melt temperatures, flow rates, channel sizes, and melt

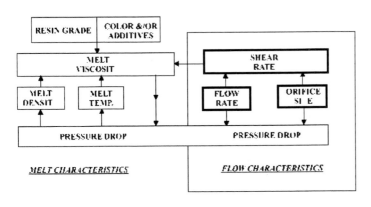

Figure 1. The interacting relationship between melt and flow characteristics.

ABS: PMMA....... 0.25	PVF2.................. 0.38	PET: PBT: FEP.... 0.60
PS: PVC: SAN.... 0.30	HDPE............... 0.50	PC: PA-6............. 0.70
LDPE: PP........... 0.35	LLDPE............. 0.60	PA-6.6................ 0.75
Note: - Taken from 'Polymer Extrusion' by Dr. Chris Rauwendaal P.218)		

Figure 2. Power Law index, [n] of generic polymers.

pressures all have influences. The interactive relationships of the above variables are illustrated in Figure 1.

A very important assumption is that channel flow surfaces and melt temperatures are the same. For well designed heads, flow surface temperatures are somewhat uniform. But often there are sections at heads and manifolds, located at mechanical joints or hard to heat areas, that have exposed bare metal surfaces which lead to heat loss. Once surface temperatures start to drop, the melt slows down. If the surface temperature drops below the melting point, the melt freezes. Either situation is a source of color change problems.

MATERIAL CHANGE CHARACTERISTICS

For material changes, problems are similar. One added factor is there is a slight difference in their "velocity profiles", as shown in Figure 3, which illustrates profiles for commonly used resins. They all have the highest velocity at the channel center, and next to zero velocity for the layer next to the wall. It holds true for irregular cross sections, except their centers are harder to define.

Velocity profiles vary according to the resin's Power Law Index, [n], (see Figure 2). ABS and acrylics, whose [n]=0.25, have flattened profiles resembling plug flow, [n]=0. Polyamide 6.6, whose [n]=0.75, has a more rounded profile similar to Newtonian flow, [n]=1. In the past, polystyrene [n]=0.30 and LDPE [n]=0.35, were used to purge out color compounds from extruder and die heads. This worked because, while not a true plug flow, they were close enough to drag more of the previous material off the flow path wall.

The following test is an attempt to simulate more plug-like flow during color/material change, by cooling the melt, then heating

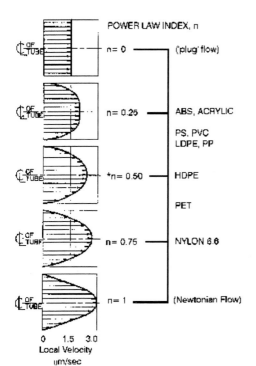

Note: * Taken from 'Flow Behavior of Thermoplastics' by A.B. Metzner, Sc.D.

Figure 3. Velocity profile of Power Law fluids (well-developed laminar, isothermal flow inside a round tube.)

the channel flow surfaces. The melt will have better cohesion and will release from the channel wall.

EQUIPMENT AND TESTING

Desired result	To alter the velocity profile by controlling surface temperature, resulting in faster color change.
Equipment	
Extruder	120 mm, 28:1 L/D; grooved feed throat with 250 HP drive
Accumulator heads	Dual 25# accumulator heads (Rear head shut-off)
Resin	HMW HDPE; 5 HLMI
Mold	22 Gal. Trash can mold [single cavity]
Colors	Black to clear (natural)

[1] Control test

Test procedure	1	Run machine with black master batch for a minimum of 500 lbs.
	2	Shut off hopper; run screw and head until empty
	3	Lower barrel & head temperature settings to 250°F for 12 hours (min.)
	4	Return temperature back to normal settings and soak for 3 hrs (approx.)
	5	Load clear material; start color change
Time - color change	Approximate 6 hours	

[2] Test with modifications

Manifold and head	Add insulation to cover top, sides and ends of manifold; also the infeed section of the accumulator heads.
Test Procedure	1) through 3) Same as above. (improve color change) 4) Set barrel zones 10°F below normal operating temp. and all manifold and head zones to max. processing temp. recommended by resin supplier. Soak for 2 to 3 hours. 5) Same as above.
Time - color change	hours - (after soak for 2 hrs) 2 hours - (after soak for 3 hrs.)
Result	Time to change color actually did improve.

CONCLUSIONS

1) Die heads are designed based on assumptions made in areas of unknown and/or unpredictable conditions. With the numbers of variables involved and the way they react under influences by one another, design accuracy just cannot be achieved for all resins and all processing conditions. This creates conditions for unbalanced and unstreamlined flow.

In attempting to streamline flow channels as much as possible, priorities have to be set when it comes to choices and compromises. Most important of course is the head functionality, the economics, etc.

2) Melt flow characteristics inside irregularly-shaped channels are complex. It is difficult to accurately analyze or predict them due to some of the melt and flow variables that are inter-related, as illustrated in Figure 1.

When designing, flow is assumed to be laminar and isothermal. In reality, neither is true; "instability" or "melt fracture" can occur in highly restricted areas or at high flow rates. Heat is transferred to and from flow surfaces according to the temperature gradients. It is being generated by melt shearing forces or from the die head heaters.

3) In order to compensate for lack of streamlined melt flow, faster color change can be achieved by controlling channel surface temperature. When surface temperature is maintained slightly higher than melt temperature, it creates a thin layer of lower viscosity film which lubricates instead of slowing down the melt stream. That reduces the chances of stagnant pockets being formed.

To accomplish this, a) Place controllable and adequate heat sources at proper locations of the die heads and manifolds, b) Cover and insulate any large areas of exposed bare metal surfaces around die heads and manifolds. This is to prevent the heat loss through free convection, and to improve response time when heat is required. It also prevents the surface temperature of any localized area inside the flow channels to fall below the temperature of melt or worse yet, below melting point. In either case, melt next to these areas will slow down, stick, or freeze onto those surfaces. When changing color, they will be the sources of streaking.

This added design feature cannot solve all color change problems for large industrial blow molding accumulator heads, but tests so far have shown good improvements on time consumed. Implementation of this is simple and relatively inexpensive, and easy to operate and control. It is worthwhile for molders who change color often to look into this different approach; time and material savings should result.

Four Color Process Compact Disc Printing: Getting as Close as Possible to Photorealism

David Scher
Autoroll Machine Corporation, Middleton, MA 01949, USA

INTRODUCTION

Compact disc printers are leaders in the plastics decorating industry. They will continue to provide innovative solutions for "Decorating in the Year 2000".

Compact disc manufacturing has experienced exponential growth that continues to push the technological envelope of how we decorate. Mass production has pushed more reliable printing with the highest picture quality imagery. CD equipment manufacturers have spent endless hours defining the pre-press parameters to help achieve the desired results. With the advent of advanced pre-press equipment for moiré detection and off-line registration, moiré free silkscreen manufacturing is now available to the compact disc decorating industry.

Looking even further beyond the boundaries of mesh, alternative decorating applications are being developed. These new methods of printing will offer even higher speeds with clarity of print that meets or exceeds the expectations of the current market. A brief history of the here and now followed by where we are heading should shed some light on one of the most aggressive plastic decorating markets.

BACKGROUND

Screen printing has faced many challenges throughout the ages. But there has never been more concern over process-color reproduction that we see today. The reason for this can be expressed in one word: photorealism, which refers to achieving near-photographic quality in a printed image. This issue has conjured up an avalanche of concerns and images of failure, particularly among compact-disc decorators.

While audio CDs usually do not require photorealism graphics, other CD applications do. For example, shops that currently screen print photo CDs are feeling pressure to achieve photorealism print results. And digital video disc (DVDs) will soon be available, promising another market for photorealism screen printing. But to achieve photorealism, screen printers have to push the limits of their process.

So why is not photorealism printing more widespread? Mostly because of production, and cost related issues. Photorealism printing is a difficult process that raises concerns about factors such as ink usage and the availability of suitable ink systems, stencil selection and durability, the availability of suitable mesh, mesh tensioning requirements, and process control. And because the printed CD must also withstand certain environmental conditions (temperatures of 185°F (85°C) with 85% relative humidity over a 21-day period) without a loss of image or disc integrity, the challenge of photorealistic screen printing is even greater.

To satisfy all these requirements and achieve a photorealistic image, you must address a number of screen-printing factors. The main areas of concern include:
- preparing artwork and separations
- selecting and preparing the screen
- creating the stencil
- selecting the ink
- setting up the press

PREPARING ARTWORK AND SEPARATIONS

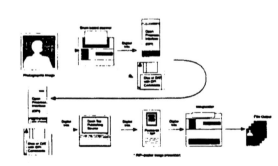

Figure 1. Separation routing.

Like other images for process-color printing, CD graphics are digitally or photographically generated, filtered, modified, and output as separations on film. You must consider the limitations of the screen-printing process as you are developing these separations because your printed image can only be as good as the separations you generate.

Photorealism graphics start with high-quality original art. The original art is scanned at approximately twice the line count of the halftone screen you want to print. The image is fine-tuned in a desktop image-editing program. Finally, the image is prepared for output as halftone separations on film. (Figure 1)

Halftones for creating photorealism images must capture as much detail as possible, so line counts generally fall in the range of 100-133 lines/in (39-52 threads/cm). The limiting factor is that no mesh is available that will hold halftones of more than 133 lines/in. (Figure 2)

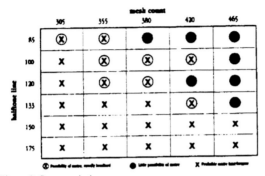

Figure 2. Screen printing process.

For smoother transitions in the midtones, elliptical halftone dots are most commonly used. And while CD printers do not use identical screen angles, photorealism separations usually feature angles that fall within the same 900 arc, just like conventional process-color separations.

The choices you make in generating separations can improve your photorealistic screen printing. Moiré is your greatest enemy, and it can be caused by a variety of factors including linear image elements, interference between halftone screens at different angles, interference between halftones and the screen mesh, improper stencil exposure, interference between the mesh and printed ink, and ink buildup. (For more information about moiré and its sources, see "Moiré Finding the Source Before You Go to Press," Parts 1 and 2, Screen Printing Magazine, Feb. and April, 1996.) And since photorealism CID printing relies on high halftone line counts coupled with high mesh counts, the potential for moiré only increases. So the precautions you take, when generating your separations, can help minimize problems on press.

A variety of devices can be used to output your separations; however, only a few offer the high level of accuracy and image density needed to produce photorealistic images. This leaves you with the two most reliable solutions: capstan and drum-based imagesetters, which use chemical-dependent films. If you have not invested in these high-end output devices, a good service bureau is your best alternative.

ALTERNATIVE SCREENING

Alternative screening techniques may reduce moiré and enhance image quality. One alternative is stochastic screening, also known as random-dot or frequency-modulated (FM) screening. This method does not rely on linear columns of dots to produce a completed image. Instead, the image is broken into varying concentrations of randomly placed, identically sized dots.

What are some of the limitations of alternative screening for CID printing? One problem is dot size. Since we are trying to achieve photorealism quality, our dots must be very small. And because the standard stochastic dot is roughly equivalent to 25% the size of a standard halftone dot, the stochastic dot for a 120- or 130-line screen ruling would be too small-no stencil could hold a dot that small, and no mesh could support the dot (Figure 3). The other obvious limitation of alternative screening is the quality of the shadows or defining edges within

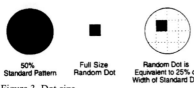

Figure 3. Dot size.

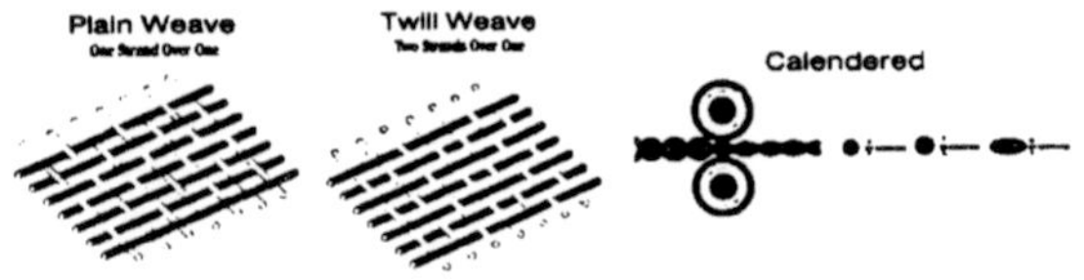

Plain Weave
One Strand Over One

Twill Weave
Two Strands Over One

Calendered

Figure 4. Comparison of available mesh sizes.

Cause	Effect
Lower off contact......................	Less image distortion
Improved ink separation.............	Faster print speeds
Faster snap off...........................	Improved image quality
Less image distortion...................	More accurate registration
Mesh openings consistent size....	Consistency of print
Less pressure on squeegees.........	Less squeegee wear
Reduced friction on mesh............	Less mesh wear
Rreduced thread diameter...........	Less ink consumption

Figure 5. High tension.

an image, which are somewhat grainy or soft in appearance. The reason for this is that color shifts are more gradual with stochastic than with standard halftones.

PREPARING THE SCREEN

The first consideration is the mesh type and thread count in use. As mentioned earlier, photorealism CID printing requires halftones with high line counts, so high thread counts are also high. Most decorators opt for plain-weave monofilament polyester mesh in counts ranging from 390-460 threads/in. (150-180 threads/cm) with a 27-micron thread diameter (Figure 4). A high thread count and small thread diameter leads to a finer printed ink deposit and, consequently, high print quality. The obvious drawback to using a small thread diameter is that it reduces the life of the mesh under normal wear. Using a plain-weave mesh is also advisable because it is less prone to leaving mesh marks and a heavy ink deposit, and will cause less print distortion than other mesh types.

Finally, you need to address mesh tension, one of the most influential prepress factors in avoiding moiré and a host of other problems on press. Among other benefits, high-tension screens help you avoid image distortion, maintain print consistency, assure accurate registration, and improve screen snap off allowing you to lower the screen's off-contact distance and operate your press at a higher speed. High tension also prolongs squeegee and screen life (Figure 5).

For realistic CID printing, high tension means being within 15-20% of the mesh manufacturer's maximum tension recommendation. Also, to ensure registration and print consistency, all screens should be tensioned to within 1-2 N/cm of one another. Printing tolerances decrease exponentially with reduced tensions.

CREATING THE STENCIL

After selecting mesh and preparing high-tension screens, it is time to turn your attention toward the stencil. Four types of stencil systems are available: indirect film, direct/indirect (a combination film/emulsion system), direct emulsion, and capillary film (also a direct stencil system). The main characteristics you try to control through emulsion coating and exposure techniques are stencil edge definition and printed ink-film thickness. The thicker the emulsion coating on the print side of your screen, the greater the ink-deposit thickness. But remember, you are trying to provide a very thin ink layer to avoid ink buildup and optimize curing.

Coating consistency is crucial when producing screens for photorealistic printing. If the emulsion thickness is not controlled, it could lead to uneven ink deposits and visual distortions such as moiré. For this reason, automatic coating devices are recommended. But if the cost of such devices is prohibitive, then maintain consistency by introducing as few variables as possible to the manual screen-coating process, including the number of personnel involved.

It is difficult to determine the correct stencil Rz value (print side surface roughness) for stencils used in photorealistic printing. On one hand, this value should be high enough to offset the problems associated with a very smooth stencil surface — cobwebbing or ink spreading beyond image edges due to static dissipation between the screen and substrate. However, if the Rz value is too high, edge definition may suffer, and ink deposit may be more difficult to control.

One key to achieving the best combination of stencil profile and surface roughness lies in the selection of the stencil material. The right dual-cure emulsion must be selected for its wet-on-wet coating properties, solids content, and print definition. The fact is that you may have to accept a little give and take between print clarity and ink transfer. In photorealistic CID printing, Rz values for direct emulsion stencils generally are measured on a Rz scale from 4-10.

Proper exposure is also critical to creating a durable and detailed stencil. Underexposure can cause stencil breakdown, which will require you to remake the entire set of screens. If an underexposed screen does not lead to breakdown, it can still affect the quality of the print, causing less detail in the shadow portions of your image. In CID printing, image areas with tonal values of 70-80% or more generally print as solid colors which means that printing one separation can degrade the other process colors in the vicinity and create what appears to be a 100% tonal area in your image. Overexposure can cause undercutting of individual halftone dots in the stencil, resulting in a loss of detail and tonal range in the highlight and shadow areas of the printed image. Since UV ink has a translucent quality, the silver disc or white underbase only intensifies the problem.

Aligning the separations with your screens is also a critical step. One successful practice is to use diecut Mylar@ masks that can be pin registered to the screen frames, allowing the film positive to be accurately positioned. However, due to the high thread counts and screen rulings common with photorealism, you must be aware that the mesh and halftone can combine to create moiré. For this reason, you may want to use a more sophisticated registration device to align the screen and film.

After applying the emulsion coating, use an exposure calculator to determine the proper exposure time. Be aware that the quality and consistency of your print depends upon each stencil being exposed using identical procedures and tolerances. If inconsistencies exist, your stencil may break down on one or more screens.

SELECTING THE INK

The inks used in CID decorating must have no ill effect on the polycarbonate discs and must support fine-detail printing. For these reasons, UV inks are the systems of choice.

Among other requirements, the ink must be viscous enough to sit on the squeegee side of the screen for 30 seconds or more without flowing through and prematurely imaging the disc. However, it must also have a low enough viscosity to flow properly and prevent trapped air bubbles and other irregularities from degrading the print quality. Flow agents are available to improve flow characteristics and reduce these concerns. To provide some color control, clear UV resins without pigmentation have been developed because UV ink colors cannot be directly modified by adding pigments. Clears do not affect the relative viscosity of the ink; rather, they enhance the translucency of the ink. Generally, though, you should avoid modifying UV inks since most are formulated for use straight out of the can.

Another qualification for the inks is that all colors must hold their hue when printed over the UV clear coating that protects the disc's reflective surface. For most photorealistic jobs, the surface of the disc is printed with a solid white underbase or "doughnut," which allows process colors to gain opacity while preventing halftones from being mirrored and forming a ghosting effect as they reflect off the CID. Most color concerns in photorealistic printing involve this underbase rather than the process-color inks.

The greatest misconception in CD decorating is that process colors are responsible for all color inaccuracies. But the white ink used in the underbase may also be responsible for subtle color variations. White inks are always based on some identifiable colored pigment, such as blue, red, yellow, or green. Each of these pigmented whites absorbs and reflects different light wavelengths, creating subtle color shifts.

SETTING UP THE PRESS

When your screens go to press, you have to assume that you have satisfied all of the prepress concerns outlined. However, the adjustments you make when setting up your press will also affect your success with photorealism. The major areas you need to be concerned with include the following:

- setting up and registering the screen

- setting off-contact distance
- setting squeegee and floodbar angles
- selecting squeegee durometer
- setting floodbar height
- setting the press's cycle speed

Your first step on press is to align the screens in the printheads. However, because both the CD and image are round, you must consider not only X and Y registration, but also the rotational alignment of your screens.

Because CDs do not offer excess surface area for printing registration marks, all screens must be manually aligned. A good practice for saving time during registration is to find a defining edge within your image to use as a master guide for aligning all colors.

Next, set your screens' off-contact distance from the CD surface at an appropriate height, which will vary depending on screen tension. Combining sufficient tension with the correct off-contact height helps assure good dot definition and minimal image distortion in the direction of the squeegee stroke by allowing the screen to snap off of the CID surface.

A coinciding issue involves setting angles of the squeegee and floodbar. In earlier CD-printing presses, the squeegee was positioned at 45° relative to the screen, while the floodbar was perpendicular to the screen. CD decorators have found that adjustments in one or both of these angles are critical to ensure that the press has the proper rolling motion to transfer ink into the mesh openings.

If squeegee and floodbar angle adjustments are not enough to achieve proper ink transfer, you can change squeegee durometer. For photorealistic CD printing, squeegee durometers ranging from 60-80 Shore A are most popular. Also consider the edge profile of your squeegee: A sharp squeegee edge leads to decreased ink transfer, while a rounder edge increases ink transfer.

If a squeegee change is not the solution, the problem could be caused by floodbar height. Normally, the floodbar moves slightly above the mesh. But testing has shown that slight contact between floodbar and screen, although slightly risky in terms of mesh damages, may help transfer ink to mesh openings.

The last concern involves the cycle speed of the press, which must be considered during screen preparation as well as during production. Why produce a set of screens with low tensions if you know that your press's top cycle speed requires very fast screen snapoff? In general, the top CD screen printed cycle their presses at around 70 discs/min.

ALTERNATIVE PRINTING

Like a fifth-generation photocopy, a screen-printed photorealistic image is not identical to the original graphic-something is lost in the translation from the original print. Attention to process control will ensure that image loss is minimized in every successive step. Although screen printing does have inherent limitations, more often than not, loss in image quality is caused by inconsistent or substandard processing techniques. If we upgrade our techniques while pushing the envelope of screen-printing technology, we will be able to deliver the next generation of CDs with even greater photorealism.

The next generation may provide an alternative to the difficulties of screen printing. Offset, Letterflex, and Flexographic, have been viewed as superior print processes due to their broad range of color replication and finer resolutions. The limitations have always existed in the slow cycle rate. Innovations in newer equipment offers continuous motion feeds that allow for CD decorating at speeds up to 100 discs per minute. (as compared to the fastest screen printer at 70 discs per minute) Problems relating to mesh such as excessive ink buildup and moiré are all but eliminated. Decorators can now approach registration and process control in the same manner that web press operators have for decades. Supplies, such as print plates, inks, and doctoring devices provide longer service with even greater photorealism.

REFERENCES

1   R. Balfour, Process Color with UV Inks: Taking Control of Your Screen Making. February, 1995.
2   M. Caza, High Tension Screens and the Quality of the Printed Image. *Screen Printing*, April, 1995.
3   J. Clarke, Control Without Confusion, St. Publications, 1986.
4   M.A. Coudray, Understanding Halftone Moiré, *Screen Printing*, October, 1991.
5   M.A. Coudray, Understanding and Controlling Halftone Dot Gain, *Screen Printing*, June 1992.
6   M.A. Coudray, Selecting a Digital Output Device, *Screen Printing*, January, 1992.
7   M.A. Coudray, Expanded Gamut Process Printing, *Screen Printing*, June, 1995.
8   M.A. Coudray, Tracking Moiré Finding the Source Before You Go to Press, Part 1, *Screen Printing*, January, 1996.
9   M.A. Coudray, Interpreting Imaged Moiré, *Screen Printing*, April, 1996.
10   M.A. Coudray, Tracking Moiré: Finding the Source Before You Go to Press, Part 2, *Screen Printing*, April, 1996.
11   S. Duccilli, C. Latscha, The Comparative Screen Fabric Guide: Mesh Selection Without the Mystery, *Screen Printing*, March, 1996.
12   G. C. Field, **Color and its Reproduction**. *GATF*. 1988.
13   T. O. Frecska, Developing ISO Standards for Process Color Screen Printing, *Screen Printing*, July, 1995.
14   GATF. Technical Guide to Scanning. 1995.
15   P. Green, **Understanding Digital Color**. *GATF.* 1995.
16   S. Hoff, How's the Water, *Screen Graphics*, May/June, 1995.
17   D. M. Hohl, Estimating Ink Deposit in Screen Printing: Improving Your Accuracy, SPTF Practical Application Bulletin.
18   M. Kennedy, Mesh Selection for Halftone Printing, *Screen Printing*, March, 1996.

Improving the Processability of Fluorescent Pigments

David A. Heyl
Day-Glo Color Corporation

Fluorescent pigments for plastic applications have presented a variety of processing challenges in the plastics industry. One such problem encountered by the end-user is plateout. The focus of this paper will therefore, aim at defining, determining the cause of, and minimizing the impact of plateout associated with fluorescent pigments — in order to improve their processability.

The term plateout has generally been referred to as "the tendency of certain pigments to form deposits on metal parts (calendar and mill rolls, embossing cylinders, extruder screws)[1] -or- "Strongly adhering deposition of pigments or other additives on machinery parts during the processing of plastics."[2] The fact that fluorescent pigments are actually dyed, polymeric resins, is the main reason that they do have a tendency to "plateout".

In general, the "plateout" material from a fluorescent color concentrate is composed of lower molecular weight fluorescent pigment carrier resin fractions, color concentrate carrier resin, and other additives and materials that are present during the molding process. This mixture of "plateout" normally appears as colored deposits on the mold face, or condenses on the blow pin assembly of blow molding equipment.

Throughout the years, efforts have been made to develop resin systems that are by nature low in plateout properties,[3] and also to develop additive packages that would additionally aid in reducing plateout.[4] It has become a relatively common practice with fluorescent color concentrate suppliers to add a small percentage (i.e., 3-8%) of silica to the masterbatch during compounding. The addition of silica materials has shown to substantially reduce the amount of plateout from fluorescent color concentrates. Due to the improvements that have occurred with different silica products, a study was undertaken to determine if any differences were apparent between fumed and precipitated silica.

For these evaluation, two different plateout testing procedures were utilized: 1) Injection mold plateout and 2) Blow-pin plateout testing. In the first testing procedure, a subjective determination is made as to the area and severity of plateout covering the mold face of an injection molding machine. The general testing procedure consists of the following:

INJECTION MOLD PLATEOUT TEST

Color concentrate is prepared and a letdown ratio of 1:20 in blow molding grade high density polyethylene is used. A 75 ton Newbury injection molding machine Model H475RS is used in this test to mold a three-step color chip of generally rectangular shape, approximately 2 1/411 by 4". Beginning with a clean mold, 3,000 grams of the letdown mixture is molded into about 275 color chip samples. After molding of the color chips is complete, the percentage area of the mold face covered by plateout is determined for each of the three steps. The intensity of the plateout as indicated by the thickness thereof is subjectively estimated for each step using a scale of 0 to 5, with zero being no plateout. These three area percentages are averaged to determine a single number evaluation of 0 to 5 injection mold plateout.

For the second test, a quantitative measurement is made of the amount of plateout material that builds-up on the blow-pin of a extrusion blow molding machine. The testing procedure consists of the following:

BLOW PIN PLATEOUT TEST

Color concentrate is prepared, and a letdown ratio of 1:10 in blow molding grade high density polyethylene is used. A Rocheleau blow molding machine Model SPB-2 is utilized in this testing procedure. The blow pin is modified to include a removable stainless steel insert on which plateout can be quantitatively determined. The amount of plateout is determined by weighing the insert before and after the blow molding of the resin. A total of 20 pounds of resin and color concentrate are blow molded, producing approximately 100 bottles. After the blow molding is complete, the blow pin insert plus the plateout is weighed and the amount of plateout can thus be determined by subtracting the original weight of the insert, giving the amount of plateout in milligrams.

The following color concentrate samples were prepared and evaluated for both blow-pin and injection mold plateout. All of the samples were produced under the same processing conditions, i.e., Killion 1 1/4", single screw extruder @ 375°F melt temperature, and 120 RPM.

Table 1 contains the different color concentrate samples and their compositions. Table 2 displays the results of the Blow pin plateout and Injection mold plateout testing procedures.

From these results it can be concluded that the use of silica, either fumed or precipitated dramatically reduces the amount of plateout that builds-up on the mold surface of an injection

Table 1. Formulations

Sample	Formula	Silica type
1	20% Fluorescent Pigment A 6% TiO$_2$ 74% LDPE	None
2	20% Fluorescent Pigment A 6% TiO$_2$ 6% Silica A 68% LDPE	Fumed
3	20% Fluorescent Pigment A 6% TiO$_2$ 6% Silica B 68% LDPE	Fumed
4	20% Fluorescent Pigment A 6% TiO$_2$ 6% Silica C 68% LDPE	Precipitated
5	20% Fluorescent Pigment A 6% TiO$_2$ 6% Silica D 68% LDPE	Precipitated

Fluorescent Pigment A	Day-Glo Color Corp. FIRE ORANGE NX-14
Silica A	Degussa Corporation TS-100
Silica B	Degussa Corporation Aerosil 200
Silica C	Degussa Corporation Sipernat 22
Silica D	Degussa Corporation Sipernat 50

Table 2. Results of testing

Sample	Blow pin plateout test, mg	Injection mold plateout test
1	2.5	1.20
2	1.8	0.40
3	1.1	0.30
4	2.2	0.55
5	2.2	0.55

molding machine. Samples 4 and 5, which are both precipitated silicas decrease the amount of injection mold plateout (Figure 1) by a factor of at least 2, whereas the fumed silicas in samples 2 and 3 decrease the injection mold plateout by impressive factors of 3x and 4x respectively (Figure 2). These results are reinforced by the quantitative analysis recorded in the

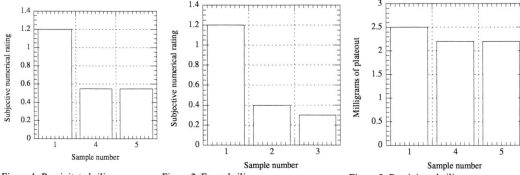

Figure 1. Precipitated silica.
Injection mold plateout.

Figure 2. Fumed silica.
Injection mold plateout

Figure 3. Precipitated silica.
Blow pin plateout.

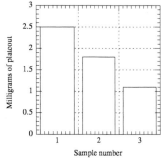

Figure 4. Fumed silica.
Blow pin plateout.

blow pin plateout results. Again, the control amount of plateout is reduced by 12% using precipitated silica (samples 4 and 5). Silica samples 2 and 3 reduced the amount of blow pin plateout by 28% and 56% respectively (Figures 3 and 4).

From these controlled test results, and also from real world molding operations where silica has been used in commercial fluorescent color concentrates, it is quite apparent that processing problems associated with plateout can be dramatically reduced by the addition of a silica product (either fumed or precipitated). Also, the use of silica will help in the dispersion of fluorescent color (i.e., less streaking) and in minimizing the appearance of the weld line on blow molded bottles.

ACKNOWLEDGMENTS

The assistance of Chris Newbacher, Andrettia Coates, Ron Laurenzi, and Connie Miceli is gratefully acknowledged.

REFERENCES

1.   L. Nass, C. Heiborger, **Encyclopedia of PVC**, *Marcel Dekker, Inc.*, New York, 1988, 594.
2.   M. Ahmed, **Coloring of Plastics**, *Van Nostrand Reinhold Company*, New York, 1979, 224.
3.   **U.S. Patent 5,094,777** (Mar. 10, 1992), T. DiPietro (Day-Glo Color Corp.).
4.   **U.S. Patent 4,820,760** (Apr. 11, 1989), M. L. Ali, J. Bateman, M. Man (Ferro Corporation).

Understanding Test Variation. A Plastics Case Study

Scott Heitzman, John Sheets

Sun Chemical Corporation, Pigments Division, Cincinnati, OH, USA

ABSTRACT

This paper explains how to use statistical tools to evaluate the reproducibility of color testing. Several pigment chemistries combined with several test methods were evaluated in hopes of establishing a test that both the statistician and color technician would endorse.

INTRODUCTION

This study was stimulated by the market's demand for tighter color specifications. We needed to quantify and possibly improve our test method reproducibility before we could realistically attempt to improve our process capability and tighten specifications.

Pigments and pigment dispersions have been and continue to be evaluated for color using a wide variety of methods. Over the years Oil Ink Tests, Latex Paint Tests, Liquid Ink Tests and PVC, Rubber, and Polyethylene Two-Roll Mill Tests have been used as Quality Control methods. We did not have extensive reproducibility data for these methods, but we felt that all of them could be improved using statistical tools.

We started reaching our goal of a quick, easy and reproducible test method for quality control of pigments for plastics by surveying our plastics color concentrate customers. The survey found a wide variety of tests being used. We selected the most prevalent methods and measured their reproducibility. With these methods selected, we chose a range of pigments to represent typical high volume pigments used in color concentrates.

STATISTICAL TOOLS REQUIRED FOR UNDERSTANDING TEST VARIATION

The most common statistic used for measuring any variation is the standard deviation.[1] Think of it as a type of index about variation — the higher the index, the more variation. Because we use samples to generate data and then calculate a standard deviation, this statistic is an estimate of the true variation index.

When we calculate a standard deviation from a set of data, all the sources of variation in the generation of the data are included in this estimate. By controlling the sources of variation, we can determine a standard deviation for a particular test method. For instance, if five different technicians using two different moisture balances generated percent moisture data over the course of 10 days using the same sample, then our standard deviation for this moisture test method would include technician and equipment differences. These combined differences make up what is generally called test reproducibility.[2] On the other hand, if only one technician measures the same sample using a single balance over a short time, then the standard deviation does not include any technician and equipment differences. This variation is called the repeatability or the precision of the test method.

For abbreviation purposes, let's define S(test) as the standard deviation due to a test method. S(test) should always be qualified to identify whether this estimate is for reproducibility, repeatability, or any other set of variation components.

The easiest and most straightforward way to calculate S(test) when you have multiple data for the same sample is to generate a standard deviation on the set of repeat tests. Typically, one uses a statistical software package or the standard deviation function in a spreadsheet to generate this value.

Testing of color pigment adds some additional nuances for consideration. For instance, to perform any type of plastics application test that measures shade and strength differences between sample and standard, one must disperse the pigment into a medium and then create a display for presentation to a spectrophotometer. Experience and special studies have proven sample preparation to be the largest source of test variation and very difficult to practically control; however, this variation can be reduced by having the standard and sample prepared side by side by the same technician, using the same equipment and testing raw materials. Each single result is then a difference generated by the same technician, equipment, and raw material. This focuses our interest in test repeatability as opposed to test reproducibility. Also, we must not forget that all pigment testing is destructive in that the exact same set of pigment particles can not be put through an application test method twice. Once the pigment has been dispersed in some medium it has been permanently changed. Thus, all repeat testing includes some "near neighbor" differences which relate to the homogeneity of the sample itself. Lastly, we have to be careful not to assume all test methods have the same precision for all pigments, especially with their differences in ease of dispersion.

Once S(test) has been calculated, how do we judge if the test is valid? How do we know when S(test) is good enough? There have been a variety of statistics used to help make this judgment. One is Percent Nominal, which is S(test) as a percent of the average result.[2] It is calculated as follows:

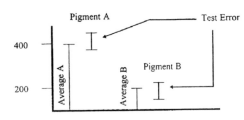

Figure 1. Conductivity.

$$\%\,Nominal = \frac{S(test)}{average}100$$

The smaller the percentage, the better the test. The problem with this method is that the amount of test variation is not always a function of the size of the results obtained.

In Figure 1, S(test) is 16.7 for Pigments A and B, with average values of 400 and 200 respectively.

Pigment A % Nominal = 16.7 / 400 * 100 = 4.2%

Pigment B % Nominal = 16.7 / 200 * 100 = 8.4%

The test variation appears to be better for Pigment A just because it has a higher average.

Another method is to analyze S(test) as a percentage of the total variation, S(total).[3] Since S(total) represents all sources of variation — raw materials, process, test method, etc. — this comparison makes good intuitive sense. Again, the smaller the percentage, the more sensitive the test is to real differences in the product. On the surface this seems like the best approach; however, it is not statistically sound! The science of statistics teaches one to compare variances (the square of the standard deviation) rather than standard deviations. One needs to think of comparing the area under the test distribution curve to the area under the total distribution curve rather than the lineal distance of the two standard deviations.

Consider comparing a room in a house to the total house. If one compares only the width of a room to the width of the entire house they may not get a valid comparison, but when one compares the square area of the room to the square area of the entire house they get a more valid ratio. This leads us to another statistic named Percent Contribution. It is calculated as follows:

$$\%\,Contribution = \frac{S(test)^2}{S(total)^2}100$$

This statistic describes the percentage of the total variation which is taken up by the variation in the test. The smaller the % Contribution, the more sensitive the test is to real differences in product samples. To be specific about the desired levels of this statistic, the following rule of thumb has been adopted:

% Contribution > 30%	Unacceptable	Too much test variation.
% Contribution < 30%	Acceptable	30% test variation
		70% raw materials & process

% Contribution < 10%   Ideal   10% test variation
                                                 90% raw materials& process

Note: The reciprocal of the % Contribution is an F-test statistic and the 30% limit equates to a significant F of 3.33.[4] This value is close to the critical F-value at 95% confidence with 8 degrees of freedom for both variances.

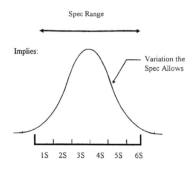

Figure 2. S(allow).

What can we do if we do not know S(total), and how do specifications fit into this statistical analysis? Since our specification range represents at least ±3 total standard deviations, then the specification range divided by 6 (±3 creates a total of 6) must equal a single standard deviation which the spec allows, called S(allow) as shown in Figure 2.

When we substitute S(allow) for S(total) in the % Contribution calculation we get a slightly different interpretation. Percent Contribution now describes the percentage of the total specification allowable variation which is taken up by the variation in the test. We now have a complete relationship between our desired % Contribution, S(test), and the specifications.

GOALS OF TEST VARIATION

Given the following specification targets, we can calculate the required S(test)s needed to achieve the 30% Contribution:

Property	Desired Specification	Required S(test) for 30% Contribution
Cielab Dl*	±0.75	0.137
Cielab Da*	±0.75	0.137
Cielab Db*	±0.75	0.137
Cielab DE*	1.3 max	0.119
Strength	±3%	0.548

CASE STUDY - TESTS TO EVALUATE AND PIGMENT FOR USE IN EVALUATION

Considering the above requirements, we decided that four test methods would be evaluated: Polyethylene Compound Test, Mix & Mold Test, Precompound and Mold Test, and Concentrate-Extrude and Mold Test.

POLYETHYLENE COMPOUND TEST

The Polyethylene Compound is a two-roll mill test. It, or some similar method, is currently used by several pigment manufacturers. Pigment is blended into pulverized and pelletized low density polyethylene. The compound is added to a heated two-roll mill and is cut and slashed by hand for 5 minutes. Swatches are cut out of the polyethylene skins and are pressed out side by side on a Carver press.

The Polyethylene Compound Test is a time-tested industry standard. It is labor intensive due to the constant cutting and slashing. Material and equipment costs are relatively low.

MIX & MOLD TEST

The Mix & Mold is an injection molding test. A premix is made using a shaker or an Osterizer. The premix is used to charge the molding machine and display chips are molded.

The Mix & Mold test was the most frequently used incoming Quality Control method for the color concentrate manufacturers. The Mix & Mold is much less labor intensive. Equipment costs are slightly higher. Multiple displays can be made with this procedure.

PRECOMPOUND AND MOLD TEST

The Precompound is a premix that is extruded and injection molded. The premix is a small amount of pulverized polyethylene that is mixed with pigment and osterized. This small master batch is then letdown in the remaining polyethylene and extruded at the end use levels. The compound is then injection molded.

The Precompound test is a three phase process, requiring three pieces of equipment. Thus, equipment expense is higher. Although not labor intensive, the Precompound test is time-consuming due to the multiple steps. Process is very similar to "Real World" conditions and should correlate well with production.

CONCENTRATE AND MOLD TEST

The Concentrate is a high level of pigment mixed with pulverized polyethylene. This preblend is then extruded on a twin-screw. The concentrate is letdown at end use levels and compounded on a single screw extruder. The compound is then injection molded to make displays.

The Concentrate test has a very high level of dispersion that simulates pigment development typical of that in an end-use application. The four phase process is very time consuming and equipment and material costs are high.

Knowing that pigments differ chemically and in dispersion, strength, particle size, color, etc., we felt it important to use a variety of pigments with different colors and chemistries. Five of our plastic grade pigments were selected for this study: phthalocyanine blue and green, staples of the organic pigment industry; an azo-based diarylide yellow because they are so prone to contamination problems; an azo-based calcium 2B as it is a "Work Horse" pigment that covers red; and a quinacridone violet because it tends to be difficult to disperse versus other organic pigments and to add a violet shade. The specific codes with pigment type are listed in the chart below.

Sun Pigment Code	Name	Index Number
264-0414	Phthalocyanine Green	Pigment Green 7
249-1284	Phthalocyanine Blue	Pigment Blue 15:3
234-0077	Calcium Red 2B	Pigment Red 48:2
274-3954	Diarylide Yellow (OT)	Pigment Yellow 14
228-5199	Quinacridone Violet	Pigment Violet 19

STUDY SCHEME

Before time and effort were spent on the test methods that utilized the molding machine, some preliminary work was required to check the reproducibility of the molding machine itself. An homogenous concentrate was created for a blue pigment sample. The pigment sample, zinc stearate, and TiO_2 were osterized three times for 30 second intervals. This premix was then added to a bag of PE resin powder and vigorously shaken. After two extrusion passes, 300 chips were continuously molded. Using the first chip as standard, 299 comparisons were made. Note the very first chips were discarded for purge considerations. Results are as follows:

Color Property	S(test) Blue - 299 Chips
Cielab DL*	0.016
Cielab Da*	0.012
Cielab Db*	0.012
Cielab DE*	0.016

It was obvious from the data presented that the molding machine chips were very repeatable and they would contribute only a small amount of the total S(test) for the entire test method.

The next step was to design the data scheme needed to obtain a good estimate of S(test) for each of the four test methods. Since we already had prior estimates of S(test) for the Polyethylene Compound Test involving a variety of products, only five comparisons of a sample to a fresh standard, each test being performed on a different day, were completed for each product. For each of the other three tests, we produced a set of chips (the number of chips was predicated by the size of the samples created) on six different days. On each day, the first chip was accepted as standard and the remaining chips were measured against it. The Mix and Mold test method created 14 chips (13 comparisons) per day for six days, giving 78 data points to analyze. Both the Precompound and Concentrate test method created 25 chips (24 comparisons) per day for six days giving 144 data points.

RESULTS AND CONCLUSIONS

Table 1 summarizes the study. S(test) was calculated using the Classical Method.

Table 1. Data

Pigment	Test method	DL* S(test)	Da* S(test)	Db* S(test)	DE* S(test)	Strength S(test)
Green 7	PE compound	0.036	0.127	0.048	0.077	0.74
	Mix & mold	0.046	0.049	0.016	0.032	0.42
	Precompound	0.018	0.040	0.013	0.030	0.18
	Concentrate	0.024	0.022	0.010	0.018	0.23
Blue 15:3	PE compound	0.085	0.027	0.065	0.099	0.58
	Mix & mold	0.043	0.028	0.035	0.027	0.31
	Precompound	0.017	0.018	0.035	0.021	0.21
	Concentrate	0.036	0.020	0.029	0.031	0.34
Red 48:2	PE compound	0.123	0.307	0.178	0.147	1.43
	Mix & mold	0.059	0.059	0.054	0.050	0.49
	Precompound	0.037	0.051	0.032	0.037	0.51
	Concentrate	0.044	0.095	0.031	0.074	0.32
Yellow 14	PE compound	0.061	0.110	0.184	0.171	0.82
	Mix & mold	0.046	0.093	0.078	0.069	0.35
	Precompound	0.046	0.093	0.118	0.125	0.36
	Concentrate	0.023	0.017	0.075	0.045	0.33
Violet 19	PE compound	0.191	0.203	0.084	0.127	1.66
	Mix & mold	0.065	0.071	0.027	0.063	0.49
	Precompound	0.056	0.063	0.043	0.048	0.67
	Concentrate	0.046	0.045	0.026	0.035	0.49

Table 1 above summarizes the data but the Cielab DE* S(test) and Strength S(test) bar charts (Figures 1 and 2) help us quickly see several interesting occurrences.

First, the Polyethylene Compound Test has the highest S(test)s across all the products tested. The Diarylide Yellow has higher S(test)s in Delta E*, but in Strength, it was just like the other colors. This may be due to yellow being the lightest color and thus more susceptible to cross contamination. The Quinacridone Violet was just the opposite. The Delta E* S(test)s are consistent across the tests but the Strength S(test)s are higher. This is thought to be due to Quinacridone Violet being a harder pigment: its dispersion differences have a larger impact. Blue and Green have low test standard deviations across the various tests.

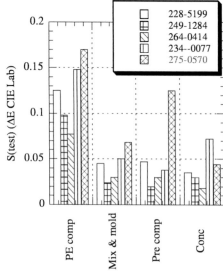

Figure 3. S(test) DE* CIE Lab.

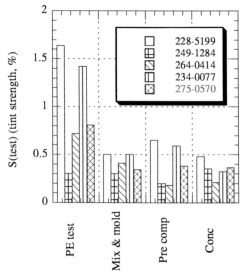

Figure 4. S(test) tint strength %.

Another observation was that the Mix & Mold test performed well. With the lower levels of dispersion, we expected results similar to that of the Polyethylene Compound Test. We also expected to see lower S(test)s from the Precompound and Concentrate Tests as compared to the Mix & Mold Test, but this was not the case. This was the most significant finding of the study, and it has permanently changed our testing direction. We could have the advantages of a more automated test without the need for expensive and time-consuming steps.

In conclusion, the Mix & Mold Test has a much lower S(test) than the Polyethylene Compound Test. It meets our minimum goal of having less than 30% test contribution of variability for strength. In fact, it is less than 10% contribution on the color components. The Mix & Mold Test is the clear choice for routine quality control testing.

IMPLEMENTATION

We have purchased and installed molding machines for our production quality control and our application and development labs. They are 28 ton machines, all with identical design, and operate well with a charge of about one pound. We are in the process of making confirmation runs and fine tuning the test method accordingly. We are also working on some correlation studies with our customers.

CONTINUED DEVELOPMENT

It's obvious this is just a step in the right direction in looking for the "perfect test." Additional development of the test is planned. Work is also planned in investigating other contributing factors such as the optimal pigment loading for each color, color strength calculations, resin and additive packages in hopes of further improvements.

ACKNOWLEDGMENTS

This study was made possible by the combined efforts of several groups within Sun Chemical Pigments Division. We wish to extend special thanks to the Plastics Group Technical Service Representatives, Dee Eichenlaub, Jim Krouse and Debra Waller for their contributions. Finally we would like to thank our customers, who have provided valuable information on their test methods and feedback on our study.

REFERENCES

1   Glossary and Tables For Statistical Quality Control, Section 3.15, *American Society for Quality Control.*
2   Richard W. Chylla, A measurement System Study: An Integral Component of Quality Management, *American Ink Maker,* June 1993.
3   A. H. Jaehn, Understanding the Effect of Sampling and Testing Error, *Tappi Journal,* August 1988.
4   Glossary and Tables For Statistical Quality Control, Table 4 Percentage Points, F Distribution, *American Society for Quality Control.*

Visual Color Matching and the Importance of Controlling External Variables

John Tasca
Techni-Cal Services, Inc.

ABSTRACT

Considering all technology available regarding color matching, visual color matching is still considered pivotal in quality control acceptance. How a sample appears will always be the under lying decision.

The function of the light booths is not simply to shine light on a sample. It is to produce a particular spectrum of light that will reflect the proper colors off the sample. If there are any shifts in the spectral output, the samples may or may not match. This is due to an effect known as Metamerism.[1] This effect will cause samples to match under one lighting condition, but not under another.

There are many items that contribute to shifting spectral outputs. These items include the age of the lamps in the light booth, the color of the inspectors shirt being worn during the inspection process, ambient light and sample size, just to name a few. We can expect a higher level of quality in color appearance by controlling these items. The only way to verify that the light booth is producing the correct spectral output is to measure the color temperature and footcandles of the booth. The color temperature reading must be ±200 degrees Kelvin and the intensity should be no less than 80 footcandles. If these readings are not within the specifications, proper corrective action should be taken to ensure the correct spectral output. This will greatly reduce color mismatches from master to sample and from light booth to light booth.

INTRODUCTION

To better understand why mismatches appear under different lighting conditions, we must first define some terms. The first is Color Temperature. The Color Temperature is a number that is used to associate a spectral output with a number and it is measured in the units of degrees Kelvin. The more commonly known color temperatures are those being used in the Light Booth. These are 2300K to simulate Horizon light, 2868K for incandescent, 4150K for Cool White Fluorescent, 6500K to simulate Average North Sky Daylight and 7500K for Simulated North Sky Daylight. To assure proper lighting condition for color matching, it has been determined that the color temperature should not vary more than ±200K.

The other term that needs defining is footcandles. This value will determine the intensity or brightness of the light.

In this case, there is a minimum value that these lamps should be. They are 80 footcandles for Horizon and 90 footcandles for the others.

Again, the color temperature is simply a number that is assigned to a particular spectrum of color and the footcandle is a unit of measurement assigned to the intensity or brightness of the light source. Controlling these two specifications is crucial to performing precision color matching and of these two, color temperature take precedence. Because this is the actual color being produced by the light booth and being reflected off the sample, if the light is not producing the color that is of the sample, it simply will not reflect that color and reflect something that is only close to it. This is where problems start to develop in visual color quality.

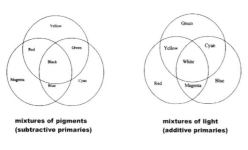

mixtures of pigments
(subtractive primaries)

mixtures of light
(additive primaries)

Figure 1. Mixtures of pigments and light.

There are many elements that can cause the color temperature to shift or to perceive the reflected color differently. As when going to a hardware store to purchase paint, the sales clerk will mix different color pigments to achieve the desired color. The same is true with light. The difference being that we do not want to change the color of the light and must take set to ensure that this "mixing of colors" does not exist (Figure 1). An example of this would be if an inspector is wearing a bright yellow shirt during color matching. As light bounces around the booth it will also bounce off the bright shirt and back into the booth. Now we have just changed the light booth color temperature.

This paper will attempt to address and offer suggestions to the most critical areas that will cause problems when color matching.

ENVIRONMENT

The conditions surrounding the light booth are important factors to consider during color matching. As you are about to see, these conditions can and should be controlled.

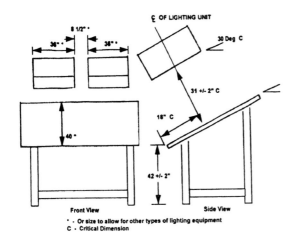

Figure 2. Light booth.

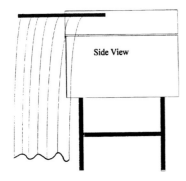

Figure 3. Curtain attachment.

Dimensions - The hanging dimensions should be set consistently with the rest of the industry. The distance of the hanging light booth to the surface of the perch must be set to 31" /- ±2". The Light Booth and the perch should be at a 30 degree angle and perpendicular to each other (Figure 2).

Surround Color - This is the color inside of the light booth and should be Munsel N9 which is a neutral gray. Over time and usage, the light booth interior may need to be repainted and brought back to its original color. In the event that the light booth is suspended, the room must be painted that same Munsel N9. This will hold true for the viewing perch as well. If a curtain is being used, this also must be as close as possible to neutral gray.

Ambient Light - Ambient light comes in a variety of color temperatures depending upon the ambient light source. Consider the red color of mercury vapor lamps or the white light from the overhead fluorescent lamps. These lights must not be mixed with the controlled light of the booth. The light booth should always be placed in an area where ambient light can be controlled. Such areas include rooms where over head lights can be turned off. Outside daylight can also be a problem. Therefore, the light booth should be away from windows that cannot block out natural daylight. If ambient light cannot be controlled externally, a booth curtain can be built or purchased to isolate the light booth from these other sources (Figure 3).

Inspector - The person performing the color matching should be aware of the potential "color mixing" that he or she may introduce to the light booth. The color of clothing may affect the overall light being reflected off the sample as discussed earlier. A simple neutral lab coat should always be available for the inspector to use in the event that bright colors are being worn. Tinted contact lenses should never be worn during inspection. These lenses slightly change the perceived color of the sample by the inspector. If this occurs, incorrect matches will develop.

Smoking should not be done in the location of the Light Booth. Tar residue sticks to the Daylight Filters, prismatic lens, reflectors and lamps. This will change the Color Temperature of the reflected light.

Maintenance - Light Booth maintenance should be performed periodically between scheduled PM's. These are in accordance with the manufacturers cleaning instructions. After disconnecting input power to the unit the following items should be cleaned:

- prismatic lens
- daylight filters
- reflectors
- cool white fluorescent lamps
- booth interior
- fan filter

These items will all contribute to the color shifting as well as the light intensity that may affect the perception of color.

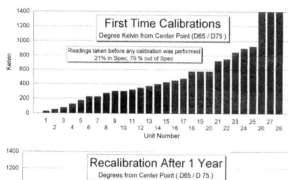

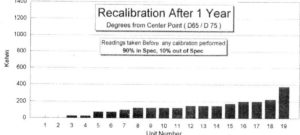

Figure 4. Calibration.

Calibration and Certification - Light booth calibration is a function that should be performed once a year or every 400 hours of daylight use. The calibration calls for the measuring and adjusting of the color temperature and footcandles. The primary reason that this work should be performed are that these units drift over time and usage. As shown in Figure 4, almost 80% of light booth that have not been calibrated, are not within tolerance. These units will cause color mismatches due to Metamerism. The calibration of the light booth verifies that the color temperature and the footcandles of each setting is within the specified tolerance of ±200 degrees Kelvin and 80-90 footcandles minimum. If this specification is not met, steps must be taken to bring the unit back to optimum performance. This may be the replacement of certain lamps, power supply adjustment or the replacement of defective components. This all should be performed by a qualified engineer with the proper equipment, traceable to NIST.

SAMPLES[2]

Sample Size - The size of the two samples being matched should be as close as possible to the same size. Larger samples have a tendency of reflecting color more vividly than smaller samples. Therefore the larger the sample the better and whenever possible, the same size will help in the evaluation.

Directional Differences - Depending upon the type of material being observed, there are different methods of viewing these sample. It is important that the samples are parallel to each other and touching. Textile material should be viewed looking into the nap. That is, when running the hand over the material and away from you, it will feel the roughest. Flat woven fabrics should be viewed with the warp yarns running up and down the examination perch. Both of these samples should be checked and matched in the Flop, Face and Traveling conditions.

Painted or glossy surfaces, plastics, leather or vinyl materials should fall into one or two categories: 1) Solid or Straight shade and 2) Metallic Colors. Solid colors should be view in the Face and Flop Positions. Plastics should be viewed with the gates facing in the same direction where the flow of resin is the same or similar. Metallic colors should be viewed in the same manner. In addition to the Face and Flop positions, Travel in very important with this material. Differences in two samples may occur at the intermediate angle as a result of the different painting technologies or spraying conditions. Always reverse the position of the master to sample when viewing. That is, change from left to right, top to bottom and verify that the matches are in all location.

Observer Differences - The sensitivity of each individual's eyes is slightly different, even for people considered to have "normal " color vision. There may be some bias toward blue o red. Blue, green and red are the primary colors of light because the eye has three types of cones (color sensors) which are sensitive to these three primary colors. This allows us to perceive color. Not all of our cones are exactly alike, making us bias to one color or another. Also, a persons eye sight changes with age. Because of these factors, color will appear sightly different to different observers.

SUMMARY

With the exception of the Observer Differences, all of these conditions can and should be controlled. If the light booth or the surround need to be painted, the paint formula is as follows: PPG 80-110 Interior Latex Flat, B-46, L-2Y, O-12. This will make one gallon of paint. Cleaning and maintenance of the reflectors and filters will keep the colors more vivid between scheduled calibration.

Annual calibration and certification is not only a requirement for ISO and QS 9000, but is good practice for maintaining precision instrumentation in which the light booth is considered. The proper color can only be reflected off a sample if that color in present in the spectral output of the light booth. Although the structural integrity of a product is most important, the appearance of the product will ALWAYS reflect the quality put into it. These steps will greatly help to achieve that reality.

REFERENCES

1   Detroit Color Council's Bulletin #3.
2   Minolta's Precision Color Communication.

Practical Analysis Techniques of Polymer Fillers by Fourier Transform Infrared Spectroscopy (FTIR)

Barbara J. Coles, Caryn J. Hall
Hauser, Inc.

ABSTRACT

The identification of polymers by FTIR is often complicated by the presence of fillers. However for kaolin clay, an FTIR analysis should be able to identify the filler and predict its concentration using a standard curve. The resulting percentage is more reliable than a simple ash, which may change the chemical composition of the filler.

INTRODUCTION

In the growing plastics industry, there is often need to identify polymer formulations. Whether the analysis is done to reproduce the material, identify another supplier, or provide insight into the cause of failure, the filler is an important aspect. Fillers are used for several reasons; to extend the amount of polymer for overall cost reduction, to add structural stability or impart specific physical characteristics to the polymer such as chemical, temperature, or flame resistance, or to add color to a polymer. Several commonly occurring fillers include: silicates, aluminum trihydrate, calcium carbonate, fiberglass, and talc. These fillers have characteristic FTIR bands which can be easily identified within a spectrum of the polymer. The amount of filler present in the formulation can be of great importance to the performance of the polymer.

THEORY

FTIR is a powerful analytical tool. Not only does it provide qualitative identification, but also quantitative information. The use of FTIR to quantify the amount of filler present in a polymer formulation should follow Beer's Law:

$$A = abc$$

where: A = absorbance, a = absorptivity (a constant specific to the material), b = thickness of sample, c = concentration.

The challenge in FTIR quantitative analysis of polymers is the thickness of the sample. The use of peak ratios standardizes the absorbance signal and eliminates the thickness variable. Attenuated total reflectance (ATR) and microscope FTIR were the two methods chosen to acquire the FTIR spectra. The filler content in the polymer was confirmed by ashing.

DESCRIPTION OF EQUIPMENT AND PROCESS

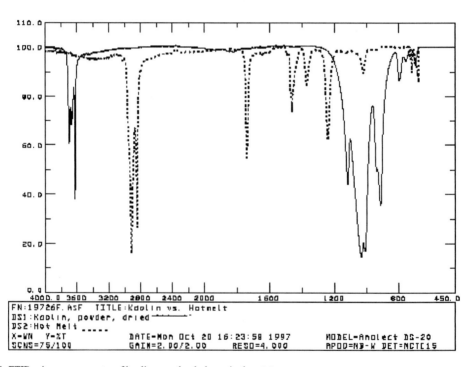

Figure 1. FTIR microscope spectra of kaolin vs. polyethylene vinyl acetate.

An Analect Diamond 20 FTIR with ATR attachment equipped with a KRS-5 45°crystal as well as a XAD-Plus Microscope attachment was used to acquire the FTIR spectra.

Kaolin powder was chosen for its peaks by microscope FTIR at 3695, 3668, 3652, 3618, 1115, 1032, 1008, 937, and 913 cm^{-1} as well as its distinctive shape above 3600 cm^{-1} and in the

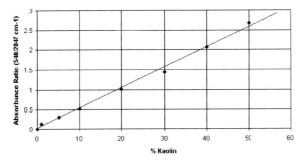

Figure 2. Standard curve for % kaolin vs. absorbance ratio (slope=0.0514, intercept=0.0118, R^2=0.995.

Table 1. Summary of results of ashing prepared samples

Calculated % kaolin in hot melt	Average % ash of samples at 500°C
0	0.03
1	0.80
5	4.19
10	8.47
20	17.68
30	26.61
40	35.59
50	44.74

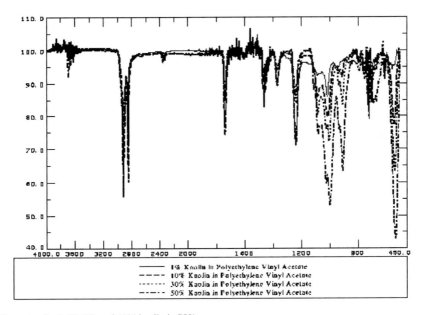

Figure 3. ATR spectra for 1, 10, 30, and 50% kaolin in PVAc.

fingerprint region. Kaolin also has a distinctive peak around 540 cm^{-1} in ATR spectra. A hot melt (polyethylene/vinyl acetate) was chosen because of the relative lack of interferences with kaolin (see Figure 1) and the ability to easily mix various amounts of filler.

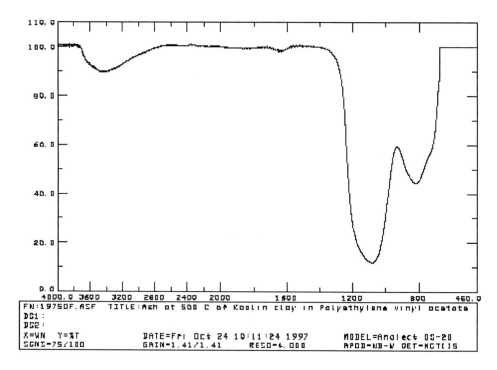

Figure 4. FTIR microscope spectrum of ashed (500°C) kaolin in PVAc.

The standards were prepared by weighing appropriate amounts of kaolin and hot melt into aluminum dishes to achieve filler percentages of 0, 1, 5, 10, 20, 30, 40, and 50%. The aluminum dishes were then heated on a hot plate to 128°C to melt the hot melt. The kaolin and hot melt were then mixed together and formed into thick films using an 8 mil draw down bar on silicone release paper. Portions of the films were cut and analyzed by both ATR and Microscope FTIR and portions were ashed at 500°C in a muffle furnace overnight and allowed to cool in a desiccator.

PRESENTATION OF DATA AND RESULTS

ATR was the most consistent tool to obtain a good correlation of peak ratios to percent filler. We obtained a standard curve with an R squared value of 0.995 using the kaolin peak at 540 cm^{-1} and the CH_2 stretch of polyethylene vinyl acetate at 2847 cm^{-1}. The standard curve is shown in Figure 2 and the overlay of the ATR spectra for 1 to 50% filler is shown in Figure 3. Summary of results of the ashing of the samples is shown in Table 1.

INTERPRETATION OF DATA

The microscope FTIR, while providing better resolution of kaolin, did not have a large enough sampling area and so was subject to small shifts in concentration of filler within the sample. ATR was not as sensitive to kaolin as microscope FTIR, but provided a larger sampling area and more consistent results.

The ashing of the samples at 500°C produced an unanticipated event. Kaolin clay holds water even when considered "bone dry." This water was liberated from the clay when it was ashed. Aluminum silicate underwent a partial transformation to aluminum oxide and silicon oxide. Figure 4 is the FTIR spectrum of the ashed material. This raises a problem with simply doing percent filler by ashing when the filler is kaolin clay; the results can be 12-20% low based on percent water and degree of conversion. Also, FTIR of the ashed material could be misinterpreted as silicates rather than aluminum silicate.

CONCLUSIONS

It is possible to predict percent kaolin by ATR examination. This method may also apply to other fillers in polymers. It is important to identify the type of filler in polymers to get an accurate picture of the polymer. However, care must be taken when a polymer is ashed then the ash analyzed by FTIR as the composition of the filler could change during the ashing process. An FTIR spectrum should be taken both before and after an ashing process.

REFERENCES

1   N. B. Colthup, L. H. Daly, S. E. Wiberley, **Introduction to Infrared and Raman Spectroscopy**. 3rd Ed. Boston: *Academic Press, Inc.*, 1990.
2   R. Gaechter, and H. Mueller, **Plastics Additives Handbook**. 2nd Ed. New York: *Hanser,* 1987.

Measuring Stabilizers in Pigmented Plastics with Near-Infrared Spectroscopy

Peter Solera, Michael Nirsberger, Noe Castillo
Ciba Specialty Chemicals Corporation

ABSTRACT

Near-infrared (NIR) spectroscopy has become a commonplace analytical tool in the plastics industry. NIR techniques have made quantification of many organic components in polymer processing easy and quick. The work presented in this paper outlines the methodology for using NIR to measure stabilizer levels in color concentrates and natural polymer formulations. Advantages and limitations are considered as well as particle size and product form.

INTRODUCTION

Near infrared (NIR) spectroscopy has been a useful analytical technique employed for many years in the pharmaceutical, cosmetic and textile industries. Its utility has been enhanced by the addition of multivariate calibration techniques. In addition, the ease of sample preparation and speed of analysis make NIR ideal for use as a quality control method.

The polymer industry has long been in search of a simple, rapid analytical method to measure stabilizer content in pigment/additive concentrates. This feasibility study was undertaken to examine the accuracy, scope and limitations of the NIR technique when applied to the measurement of traditional light stabilizers in color concentrates.

OBJECTIVE

The work described in this paper was designed as a feasibility study. When compared with large calibration sets typical of analytical methods development for routine quality control testing, the small number of data sets here pale in comparison. Rigorous laboratory preparations would be necessary before reducing the NIR analyses discussed in this work to viable QC test methods.

In this study, we set out to investigate the accuracy of NIR at stabilizer loading levels around 10% in both natural polymer and in color concentrates containing approximately 10% pigment.

Additional areas of investigation included:

- pigment interference with hindered amines (HALS) and with ultraviolet absorbers (UVA)
- compositions containing more than one light stabilizer
- the effect of product form and particle size on accuracy of method
- evaluation of NIR spectroscopy in a range of polymers.

EXPERIMENTAL METHODS

Samples were prepared by dry blending polymer, additives and dry pigment, extruding the dry mixture in a single screw extruder and pelletizing. All pellet samples were dried in an oven prior to NIR analysis. Powder samples were prepared from the pellets in a cryogenic grinder. Plaques were compression molded. NIR measurements were obtained by measuring eight separate samples for each formulation. Values reported in this paper are the average of results for each formulation. Expected additive concentrations were based on the assumption that gravimetric preparation of the formulations was accurate to ±0.005 g in 100 g. Primary analysis was performed only on samples reported in Section 3.

Extruder	Superior/MPM 1" Single Screw Extruder; 24:1 L/D
Screen pack:	20-60-20
Extrusion conditions:	

Zone 1	Zone 2	Zone 3	Die
425°F	450°F	475°F	475°F

Extruded with hopper open to air at 100 RPM

Screw design	General purpose
Compression Molder	Wabash Compression Molder
Thickness	60 mil plaques
Pressure	3 minutes low pressure @ 300 PSI
	3 minutes high pressure @ 3000 PSI
	3 minutes cooling time (ASTM D 1928)
Grinder	Bauermeister Universal Laboratory Mill
Mill speed	Set point 350
	Resulting speed 12,000 rpm
Batch size	500 grams
Grinding time	20 minutes with liquid nitrogen cooling
	Samples were frozen @ -40°C before grinding

Drying
Temperature   60°C
Time   4 hours
Air   500 air changes per hour

NEAR IR MEASUREMENTS

The NIR spectra of the polymer samples were obtained using a Perstorp NIR Systems Model 5000 Spectrophotometer equipped with a Rapid Resin Analyzer. The raw spectral measurements were made in the reflectance mode from 1100 to 2500 nm. Between 150-200 grams of sample were poured into a pellet sample cell and inserted into the instrument. The sample cell was slowly moved through the NIR beam as the spectra were collected. Approximately 60 cm^2 of sample were presented to the spectrometer and 8 scans per formulation recorded. The raw sample spectra were enhanced by conversion to second derivative spectra. A partial least squares (PLS) regression was performed on the second derivative spectra to generate the calibration model. The regression was generally carried out over the range 1100 - 1900 nm or at specific narrower ranges to take advantage of unique sample or analyte spectral absorption features.

PRIMARY ANALYTICAL METHOD - TOTAL BASICITY

Samples analyzed for HALS content *via* total basicity were refluxed for 15 minutes in 100 ml toluene. After refluxing, solutions were allowed to cool. Approximately 100 ml of a 1:1 mixture on acetonitrile and chloroform were added followed by 20 ml of glacial acetic acid. Samples were then titrated with a standardized 0.1 N perchloric acid solution.

RESULTS

1 - BUILDING CALIBRATION SETS

For our original work, we used a scan region of 1100 - 2500 nanometers (nm) for HALS 1 and performed the PLS regression over a range of 1100 - 1900 nm. Samples containing 9%, 10%, 11 % HALS 1 in natural LLDPE were prepared, and several samples were analyzed at each concentration. The results are presented in Table 1 and Figure 1.

Table 1. NIR results for HALS 1 in natural LLDPE

Sample ID	Expected level HALS 1	NIR measured level	Range of variation
N-45	0.00	0.034	±0.19
N-41	9.00	9.100	±0.20
N-42	10.00	10.003	±0.12
N-43	11.00	10.789	±0.24

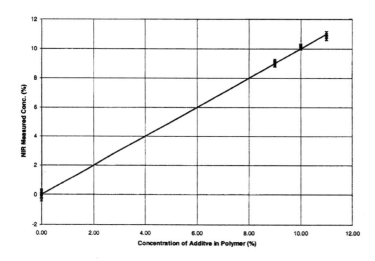

Figure 1: Natural LLDPE with HALS 1.

A similar set of samples was prepared in a 10% Pigment Red 144 concentrate in LLDPE to observe the effect of the pigment on the calibration curve. Once again, the calibration curve with the limited data set shows excellent agreement with expected values (Figure 2).

These two data sets demonstrate that reasonably good experimental fit can be achieved in both natural and pigmented polymer/additive concentrates for high molecular weight HALS without extensive sample preparation.

2 - ACCURACY IN A NARROW CONCENTRATION RANGE

The accuracy of the method was examined with a series of samples containing HALS 2 at various levels between 10.5% and 11.00%. We refined our original method by selecting certain regions of the NIR which varied with changing composition of the target additives.

When selecting a spectral region for quantification, it is helpful to examine samples of the polymer with and without the target additive. Samples should be fully formulated with other expected components

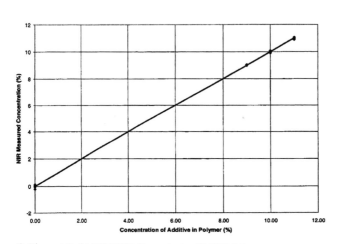

Figure 2: Pigment Red 144 LLDPE Concentrate with HALS 1.

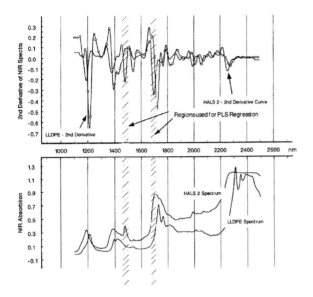

Figure 3: Bottom - NIR spectra of LLDPE and HALS 2; Top - second derivative of NIR spectra.

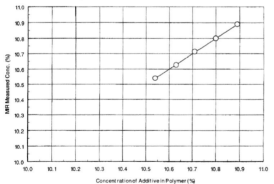

Figure 4: Calibration set for narrow concentration range of HALS 2 in natural LLDPE.

such as pigments. In addition, a spectrum of the neat target additive may provide additional insight regarding relevant differences in the component spectra.

For this work, the spectral regions used for quantification were 1465-1500 nm and 1670-1700 nm based on the contributions of the additive compared with the polymer (see Figure 3). The regions were selected to maximize the difference between the background components (polymer, pigment) and the target additive (HALS 2).

Once again, surprisingly accurate values were obtained with a relatively small calibration set compared to typical spectroscopy methodology. The PLS regression was fitted by three factors to an R value of 0.9999 (Figure 4).

Despite the good fit illustrated above, great care must be exercised when trying to set up calibration curves. The calibration set was extended with a sixth sample prepared on another extruder with a different diameter die which gave a slightly larger pellet size. When this spectrum was included in the calibration set, it resulted in a regression curve with less accuracy compared with values without N 1236 (Table 2).

3 - VALIDATION OF NIR METHOD WITH PRIMARY ANALYSIS

It is important to verify the actual levels of the additive either by careful control during loading or with an accurate, reliable primary analytical method.

Table 2. Comparison of calibration sets for HALS 2 in natural LLDPE

Sample ID	Expected level HALS 2	NIR measured levels (from PLS without N 1236)	NIR measured levels (from PLS including N 1236)
N 1236	10.45		10.579
N 1222	10.54	10.574	10.654
N 1223	10.63	10.636	10.709
N 1224	10.71	10.652	10.716
N 1225	10.80	10.827	10.573
N 1226	10.89	10.881	10.789

Table 3. Comparison of primary analytical method vs. NIR measured concentrations of HALS 3 in natural LLDPE

Sample ID	Expected Level HALS 3, %	Primary analytical method (total base titration), %	NIR measured levels, %
N-73	0.00	0.00	-0.07
N-69	9.00	8.90	9.062
N-70	10.00	10.00	10.103
N-71	11.00	11.10	10.913

Table 4. Comparison of primary analytical method vs. NIR measured concentrations of HALS 3 in Pigment Red 144 LLDPE color concentrate.

Sample ID	Expected Level HALS 3, %	Primary analytical method (total base titration), %	NIR measured levels, %
N-72	0.00	0.10	-0.004
N-66	9.00	8.40	8.852
N-67	10.00	9.70	10.059
N-68	11.00	10.60	11.029

Pigment : Red 144 at 10%

In the following examples, a primary analytical test method was used to measure HALS 3 concentrations in LLDPE natural and color concentrate. The primary method utilized in this evaluation was a total base titration of the extracted hindered amine. A comparison of the expected values, primary analytical results and NIR measured values are presented in Tables 3 & 4.

Table 5. HALS 1/UVA 1 compositions

Sample ID	HALS 1, %	UVA 1, %	Pigment Blue 15:1, %	PP, %	Base stabilization, %
A	5.0	4.0	10	80.9	0.1
B	5.0	5.0	10	79.9	0.1
C	5.0	6.0	10	78.9	0.1
X	4.0	5.0	10	80.9	0.1
Y	5.0	5.0	10	79.9	0.1
Z	6.0	5.0	10	78.9	0.1

Table 6. Measured values for UVA 1 using PLS regression equation from Figure 5

Sample ID	Expected UVA 1 concentration, %	NIR measured UVA 1 concentration, %
X	5.0	6.05
Y	5.0	5.23
Z	5.0	4.31

Table 7. Measured values for HALS 1 using PLS regression equation from Figure 6

Sample ID	Expected HALS 1 concentration, %	NIR measured HALS 1 concentration, %
A	5.0	4.93
B	5.0	5.07
C	5.0	5.01

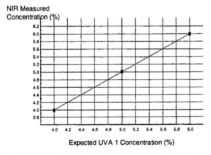

Figure 5. Calibration curve for UVA 1 in PP using Samples A, B, C.

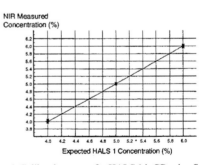

Figure 6. Calibration curve for HALS 1 in PP using Samples X, Y, Z.

In the natural samples, there was fairly close agreement between the NIR method and primary analysis. However, in the pigment concentrate, the primary method seemed to give values less than theoretical - about 0.3% to 0.6% - while the NIR method was fairly accurate. This discrepancy between theoretical levels and reported values has traditionally been difficult to avoid when intensive laboratory preparation is required.

4 - MULTIPLE ADDITIVE COMPONENTS

HALS & UV Absorber

It is well documented in the literature that NIR calibration curves for multi-component formulations should be assembled by scanning standards which contain all of the components varying the level of only one component in each sample. The relative contributions of each of the components may vary considerably as is illustrated in the following test data.

Two small calibration sets were prepared with HALS 1 and UVA 1 in a blue pigmented polypropylene concentrate. The compositions are listed below (Table 5).

After scanning the samples, PLS regression and curve fitting, two calibration curves were prepared (Figures 5 and 6). Eight samples were analyzed at each loading level and fit to a PLS regression. The PLS regression equation derived for Figure 5 was used to "predict" the values of UVA 1 in Samples X, Y, Z. The results are inaccurate compared with expected values and, in fact, suggest a negative correlation between loaded values of HALS 1 and predicted values of UVA 1 (Table 6).

In contrast, the same exercise was performed for Samples A, B, C using the PLS regression equation from Figure 6. The "predicted" values for HALS 1 were very close to expected (Table 7).

This difference in the accuracy of prediction is probably an artifact of the relative contributions of each additive in the examined spectral region. It is this type of inconsistency in the predictive models which make it imperative that the calibration data sets be as complete as possible. More data points are needed to property construct an accurate PLS regression equation for UVA 1.

Combination of Hindered Amines

HALS 3 is a blend of HALS 1 and another high molecular weight hindered amine. Quantification of this additive system can be performed by treating it as a single component as was done in Section 3 - Validation of NIR Method with Primary Analysis. Alternatively, a system composed of individual components can be analyzed by measuring the contribution of each component separately as in the previous section. However, as previously mentioned, to generate meaningful data on the individual

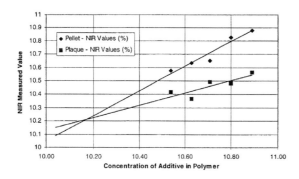

Figure 7. Product form - pellets vs. plaque.

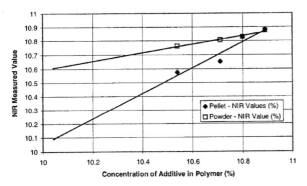

Figure 8. Product form - pellets vs. powder.

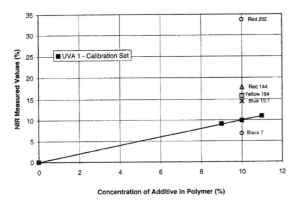

Figure 9. Effect of pigment on estimate of UVA 1.

components, calibration curves must be constructed with samples containing varying contents of each of the components. This will allow the multivariate statistical method to effectively assess the relative contributions from each of the components.

5 - EFFECT OF PRODUCT FORM

The effect of product form was examined by taking the pellet samples used in Section 2 and making plaques and powder. The powder samples were fabricated in a cryogenic grinder, and the plaques were compression molded.

The alternate samples forms were read using the same scan and PLS regression regions as in Section 2. The calibration curve from Section 2 was used to predict the values.

Figures 7 and 8 clearly show the HALS 2 level in the plaques are underestimated while the additive concentrations in the powder samples are overestimated. Since both the plaques and the powder samples were produced from the original pellets, it is not likely that HALS 2 levels would be depleted in the plaques and yet show elevated levels in the powder. These results are most likely due to differences in scattering of reflectance beam between the product forms.

It is evident from Figures 7 and 8 that product form plays a key role in obtaining accurate data. Thus, calibration curves should be generated using the appropriate product form and particle size.

6 - EFFECT OF PIGMENTS ON ESTIMATIONS FROM STANDARD CALIBRATION SET

One of the features we observed early during the collection of the NIR spectra was the apparent lack of interference from most of the pigments we examined in the NIR regions specific to HALS and UVA absorbance. We, therefore, made a calibration curve and regression equation

Table 8. NIR calibration curve - HALS 3 in ABS + 10% Pigment Red 144

Sample ID	Expected HALS 3 concentration, %	NIR measured HALS 3 concentration, %
N-1235	0	0.001
N-1205	9	9.007
N-1206	10	9.999
N-1207	11	10.993

Table 9. NIR calibration curve - HALS 3 in polyamide + 10% Pigment Red 202.

Sample ID	Expected HALS 3 concentration, %	NIR measured HALS 3 concentration, %
N-1231	0	0.002
N-1202	9	9.049
N-1203	10	9.956
N-1204	11	10.992

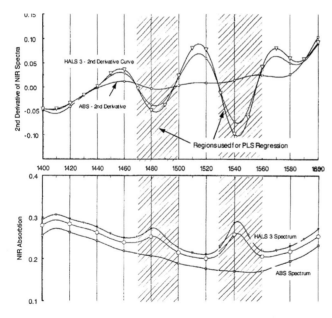

Figure 10. ABS NIR spectra and 2nd derivative curves.

for UVA 1 in natural polypropylene and attempted to "predict" corresponding UVA 1 levels in 10% pigment concentrates. As expected, this technique proved to be an inaccurate method for determining additive concentrations even though the comparative NIR spectra for the individual components would suggest little interference from the pigments.

Figure 9 summarizes the "predicted" values for UVA 1 loaded at 10% into concentrates containing carbon black, CPC blue, bismuth vanadate, azo condensation and quinacridone pigments. The NIR measured values range from 6.7% to 35% versus the expected 10%.

7 - POLYAMIDE AND ABS

Calibration curves were generated for both polyamide and ABS to investigate the capability of the method in non-olefin polymers. Tables 8 and 9 demonstrate good agreement with theoretical values can be obtained in these polymer color concentrates.

CONCLUSIONS

While the methodology employed here was not rigorous analytical technique, the results are impressive based on their accuracy and simplicity. It was apparent when examining the ability of NIR spectroscopy to predict stabilizer levels that an extensive set of calibration data is required for best results. However, it was also noted that with little work a reasonably good calibration curve can be obtained. The use of a primary analytical method is typically recommended when developing NIR methods. Nevertheless, in-polymer analysis of stabilizers can be somewhat difficult and lead to inconsistency in the calibration set. When developing NIR QC methods for polymer concentrates, it may be appropriate to use additive loading level as the best indicator for the expected measured values. This work also verifies the necessity of building standard calibration curves with samples containing all of the components expected in actual production materials. A "universal" calibration curve is probably not possible for most applications. Recognizing that this feasibility study was very limited in scope, the method appears to be useful across several different polymer types including polyolefins, polyamide and styrenic elastomer blends. Accuracy and simplicity do not often appear in the same sentence when describing analytical technique. However, multivariate calibration methods make it easier to get accurate results with NIR, and the minimal sample preparation will continue to be a very attractive feature for this analytical method.

ACKNOWLEDGMENTS

The authors would like to extend their appreciation to J. Osmundsen who helped prepare samples for this study, and to Dr. K. Ng for his guidance.

REFERENCES

1   Recent Advances in Near-infrared Reflectance Spectroscopy, *Applied Spectroscopy Reviews*, **27**(4), 325-383 (1992).
2   Near-infrared Spectroscopy of Synthetic Polymers, *Applied Spectroscopy Reviews*, **26**(4), 277-339 (1991).
3   Comparison of Mid-IR with NIR in Polymer Analysis, *Applied Spectroscopy Reviews*, **28**(3), 231-284 (1993).
4   J. W. Hall, D. E. Grzybowski, S. L. Monfre, Analysis of polymer pellets obtained from two extruders using near infrared spectroscopy, *J. Near Infrared Spectrosc.*, **1**, 55-62 (1993).

APPENDIX 1

Pigments

Pigment Red 144	azo condensation
Pigment Blue 15:1	Cu phthalocyanine blue
Pigment Black 7	carbon black
Pigment Yellow 184	bismuth vanadate
Pigment Red 202	quinacridone

Multi-Angle Spectrophotometers for Metallic, Pearlescent, and Special Effects Colors

Brian D. Teunis
X-Rite, Inc., 3100 44th Street SW, Grandville, MI 49418, USA

INTRODUCTION

The matching and quality control of color components has always been an ever challenging task. With the introduction of higher quality standards and consciousness it has become even more critical to accurately measure and reproduce color. Added to the quality issue has been the introduction of new special effects colors that change appearance with viewing angle. The use of the special effects along with metallic and pearlescent colors has generated a need for an instrumental means of quantifying these effects. This has been especially true in the automotive exterior colors.

Designers have used these special effects in new and innovative designs which has forced quality engineers to search for more consistent and accurate means of quantifying color in the manufacturing process. When evaluating exterior automotive color differences with instrumentation there are a number of variables that need to be considered, most importantly is instrument geometry. Other areas of focus are color standards, paint technologies, part configuration, part orientation and, of course, visual comparison.

While older existing instrument geometries such as diffuse/8, commonly known as sphere, and 0/45, can give some indication as to what kind of color difference exist, neither provides the correlation to visual assessment nor correlation to process parameters needed to make adjustments. Utilizing recent technology, one can now accurately monitor and control automotive colors with the use of a multi-angle spectrophotometer.

COLORS

The colors found in industry today can be grouped into four main categories of finishes; solid or straight shade colors, metallic colors, pearlescent colors and special effect colors. Any straight shade color is a finish that is solid in color across the surface. Traditional colors like white, black and solid red are examples of the straight shade colors. Metallic finishes are those colors that have metallic flake (usually aluminum) dispersed throughout the finish so that they may exhibit a "flop-like" appearance of flashy highlights to saturated deep tones. Pearlescent colors have mica particles throughout the coating not only creating highlights, but also various color changes, known as hue shifts. A special effect color can be classified as any color that does not fall into one of the other three categories. Some special effects are created by pigmenting the metallic or mica flake causing hue shifts at different angles of view. The combinations appear endless as clear tints are applied and transparent pigmentation is added to clear coats over pigmented metallic or mica flake. The effects are that the color or appearance of the color is totally dependent on the angle of reflection or the viewing angle of the observer. Add to these colors the appearance effects of surface characteristics such as orange peel, distinctness of image, DOI, and gloss and you have created a quality engineer's nightmare. For the purpose of this paper we will only discuss the control and measurement of the color issues.

What is the reason for all these different colors? Consumer demand for new ways of decorating manufactured objects such as automobiles is the primary driving force. Product differentiation by design to appeal to consumers also pushes the need to innovate and develop unique color opportunities. Automotive body styles also can lend themselves well to certain color effects. Today's automotive colors, including straight shade colors, have inherent characteristics that when applied to different body styles are exhibited at different magnitudes. For example, a vehicle with a metallic-flake maroon finish will appear as a bright flashy red along highlight lines while the same vehicle becomes a deep maroon when the metallic is eliminated by viewing at the flop angle. In the case of most metallic flake colors, the shift is mainly a lightness/darkness shift. Some straight shade colors can exhibit a similar effect, however the shift in lightness/darkness is very slight compared to a metallic based finish. This shift is typically a result of the combination of surface attributes, body style, the paint itself as well as any, how much and the type of clearcoat applied.

With the introduction of pearlescent and effect pigments, automotive designers are now able to have the vehicle actually change color when viewed at different angles. All this begins to create some challenges as to how these finishes can be reproduced in a production type atmosphere. It is important to note that the body style plays a very important role in how the different components of a vehicle must come together without a visual mismatch. If a mismatch occurs at the assembly line, it is too late to correct the color. The time, labor, materials and processing time has already been invested only to produce a part or component that does not match the other parts or body components.

The goal is quickly defined as being able to assemble a vehicle that contains a painted body and several supplied components and parts without a color mismatch issue.

Before we can start to investigate the color match issues in production, .
mine how we are going to quantify the color differences that exist. To assemble a .
may have several line to line color match areas and multiple components supplied by mun.,
vendors, painted with differing paint technologies supplied by different paint suppliers, without a single color mismatch issue, is a challenge. To do this on a day in and day out basis, is
not a challenge that can cost effectively be handled by subjective means. It must be handled
with objective quantification means of measurement so that decisions to change or alter materials and processes are made based on facts, the numbers.

If we consider the final judge of color to be the consumer with the majority of critical
viewing being done outside or on the showroom floor, then the measurement systems must be
able to simulate these conditions. As the typical observer approaches the vehicle and studies
the vehicles styling and color(s), they begin to examine the overall appearance from front to
back, top to bottom, side to side. For this reason the instrument must "see" the color from
multiple angles, evaluate differences under differing lighting conditions and must be able to
track measurement location on the vehicle or part. Most importantly, the instrumentation used
must correlate to visual evaluation.

When considering an instrument for color measurement three main geometries of color
instrumentation need to be reviewed. These are the 0/45, or 45/0 sphere (d/8) and multi-angle.
Each geometry is designed to measure specific samples for color of "color and appearance"
attributes.

Instrument geometries usually reference the illumination and detection directions in
their name. The illumination direction is specified first, followed by the detection or pick-up
angle. For example, a 0/45 instrument geometry describes illumination at 0 degrees, perpendicular to the sample plane, and the detectors or pick-up at 45 degrees to the plane of illumination. The specular energy, or 1st surface reflectance, will be reflected back at the light source
and therefore naturally "excludes" the specular energy of the sample from the measurement.

0/45 GEOMETRY

These instruments do not collect or incorporate the specular reflectance of a sample into the
measurement. This measurement is typical of the view the human eye may have of a sample
by excluding the specular energy. For example, when viewing a high gloss type of sample you
can usually determine a point where the reflection of the light is "mirrored" or reflected directly back at your eyes. This makes it difficult, as well as uncomfortable, to see any detail of
the color of the sample. By tilting the sample either away from you or towards you, you can

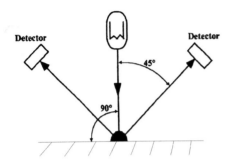

Figure 1. 0°/45° geometry.

exclude the glare. For this reason specular excluded instruments will record color differences much the same way we visually compare samples.

These type of instruments perform well in quality control situations where standards and samples have similar textures and gloss levels; however they only evaluate color at a single angle. Therefore these instruments do not give the full detail of the appearance of a color at multiple viewing angles as would be required with metallic pearlescents or special effects colors.

SPHERICAL (DIFFUSE/8) GEOMETRY

The sphere geometry or diffuse/8 geometry instruments diffusely illuminate the sample and detect the energy collected at 8 degrees off the perpendicular axis to the sample. The diffuse illumination is accomplished by illuminating a white sphere that in turn illuminates the sample from all angles, hence diffuse illumination. When using the sphere type instruments the choice can be made to either include or exclude the specular component with the measurement.

When comparing two samples of identical pigmentation, one matte finish and the other high gloss, there appears to be two different colors due to the different surface characteristics.

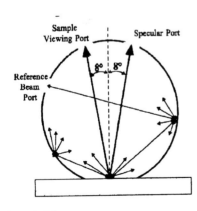

Figure 2. Sphere geometry.

Samples that have a high gloss or highly directional type surfaces will redirect the "gloss" or specular component in one general direction. Therefore under normal viewing conditions, this makes the high gloss sample appear dark when compared to the matte sample. If measured with a 0/45 or sphere excluded measurement the results would be nearly the same.

Conversely, if those two samples are measured with a sphere instrument with the specular component included, the instrument results will be very similar. By including all of the reflection from the sample, the instrument cannot detect differences in samples due to texture or gloss and will identify the color with the same pigmentation, as being the same. Even though

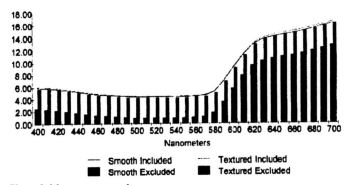

Figure 3. Measurement results.

the samples appearance is different the color of the samples is identical, due to the same pigmentation, and the specular included measurement will identify it as such. The advantage of an instrument operated in the included mode is that it measures the overall "pigmentation" of a sample. This is why many color formulators in the paint, plastic and textile industries will use the sphere included measurements to determine the proper pigmentation for a customer sample.

The advantage of a sphere instrument over 0/45 is that most sphere instruments can be operated in the included and excluded modes. In quality control of many materials we can first determine if a given sample looks different by evaluating the excluded data. If different, we can then examine the included date to see if the "formula" needs to be adjusted. The advantages of a sphere instrument have many applications in multiple different applications within the paint, plastics or textiles industries. But unfortunately the sphere geometry is inadequate in determining color differences in metallic, pearlescent or special effects colors.

MULTI-ANGLE GEOMETRY

The multi-angle spectrophotometer is able to discern color differences at different angles of reflection. Because the metallic and mica flakes and special effects colors actually change in color as the viewing angle changes the multi-angle spectrophotometer is utilized. The reference here is for the instrument with 45 degree angle of illumination to the sample plane and detection angles of 15, 25, 45, 75 and 110. These angles are considered aspecular and are measured from the specular angles. This type of instrumentation can provide valuable objective, numerical information needed to understand how color is changed as a raw material, paint process or physical parameters are changed in a production setting.

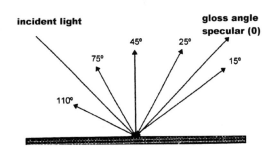

Figure 4. Multi-angle measurements.

COLOR STANDARDS AND SAMPLES

A color measurement instrument can be used as a tool to record the "lifecycle" of any color. The goal is to maintain the integrity of a color from design through implementation to production to final assembly and delivery. In the automotive field a color begins life at the design table, where a color or concept of colors is determined that will enhance a specific car or line. This would be an excellent point to capture a color measurement, because chances are the color may never be the same again.

The "color development" then continues at the paint companies. As the paint companies compete for the assembly plant business they submit sample batches to be tested, retested and tested some more at pilot operations around the world. Beside color adhesion, durability, DOI, gloss, orange peel and any other number of performance factors are studied. When the studies are complete and a color is proven "feasible" a designated paint company is then challenged to make hundreds (even thousands) of sample panels. But first, a measurement should be taken to insure these are the proper color.

Sample panels are produced and the total population of standards are evaluated, eliminating those standards with scratches, dirt, too much color variation or not close enough to the desired target color. What remains is a sample set of the "best" color standard panels. Quantify at this point can vary depending on how many were originally produced. This sample set may or may not represent the mean of the total standards population, however it will have its own color variation within the subset. To find what that actual variation may be requires measurements be taken. These standards panels are further categorized into subsets with terms like reference standard, master standard and working standard.

In the measured procedure for standards, the cars or components or parts all must have a repeatable presentation technique to the instrument. This requires detailed instructions and a proven methodology. There are several ways to approach the measurement of samples. One is to fixture the sample to ensure that the same place is measured time and again for each sample. The second is to average a given number of measurements. The third option is to use a statistical method of measurement that analyzes the measurement and accepts or rejects samples from a predefined statistical basis.

In general terms, master standards are used to check working standards and should be maintained and tracked in a laboratory environment while working standards are to be used to check production arts. A problem that exists is the range of the master standards may be approaching the allowed limits of visual acceptability. If each paint company, assembly plant and parts supplier considered their standard the target and added to the variation amongst standards is their production variation then the total variation among parts from different suppliers may be larger than the visually allowable tolerance. The challenge is to reduce variation throughout the process. Much information can be learned by comparing instrumental data collected at this point with the visual evaluations.

Reduction in standard variation can be investigated by using the multi-angle spectrophotometer and one of the following measurement methods. The first is to have a single standard that paint suppliers, assembly plants and other part suppliers all can measure. They each measure the one standard, usually fixtured, and retain the numerical value of that standard measured by their multi-angle spectrophotometer.

The second method would be to distribute "identical" standards to each supplier and paint company, this is dependent upon enough of the standards measuring as identical. Each manufacturing location then maintains that standard under laboratory conditions and utilizing the measured value from that standard as their production or working standard.

The third technique would be to have each standard hitched to a "Reference Standard". Through this type of program the physical standard that is distributed carries with it offset data that gives the position of that particular standard panel in color space as compared to the reference standard.

A fourth option would be to have a purely digital standard. A master multi-angle color spectrophotometer is defined and its color values of the standard are passed digitally to any portion of the color communication chain. In the option the standard variation is the variation of inter- instrument agreement between the multi-angle spectrophotometers in the color communication chain.

All of these methods can contribute to the reduction in variation amongst standards and is important to note that each may have significant benefits depending upon the organization and how it chooses to resolve color difference issues.

PROCESS VARIATION

During the manufacturing process numerous situations and variation can arise to cause variation and deviation from the accepted standard color. These process variations need to be determined and corrected to ensure the final product proper color match. The multi-angle spectrophotometer can be used to determine and quantify process variation.

Some factors that may arise include whether the match is being made to solvent or water based material. In certain instances the component cannot be painted in a water based system as is the body of the automobile. This can be a tough issue if the standard is also not created in the same system and has never been tested for that type of technology compatibility.

Other factors that must be accounted for are assembly plant situations. In some low running colors there may be paint degradation that require smaller batches and refreshed more frequently leading to batch differences being more prevalent than in higher usage colors. Certain colors will drift significantly during the curing process and should be monitored. Paint application equipment may affect color and should be monitored. Paint film build combined with multi-angle color analysis can provide trouble shooting information on off-color issues, but only if monitored and measured. This can be especially helpful in application issues in comparing side to side, front to rear, horizontal to vertical automotive differences. These issues may require offset standards to satisfy the tolerances.

In many cases where color differences exist, it may be necessary to evaluate the magnitude and the nature of the color shift. The use of delta E value alone will not provide any directional color information. But information such as hue versus angle of measurement can help to provide the insight to resolve particular color issues. With all the processing steps involved and process variable influence it is necessary that the measurement instrument offer the ability to provide selectable information to assist in identifying problem areas.

Additional concerns of the component supplier who provides pre-colored or molded in color parts can be resolved using the multi-angle spectrophotometer. If they are dealing with a part that has extruded or molded in color then the issues are compounded by the fact that plastics tend to have different pigment formulations, different metallic flake, different mica and pearlescent pigment characteristics all due to different coloring properties. Plastic parts also have different appearance attributes that can be altered by a unique set of processing variables. One of the most notable factors is that the process tends to lay all the metallic flake in one direction giving the part a significant "flop" type characteristic. The multi-angle spectrophotometer provides a measurement tool to characterize and quantify these effects to better control the process.

VISUAL STANDARDS

As with all these applications it is vital that meaningful data be gathered by which process can be fine tuned. As with the offset standard for paint manufacturers, it is sometimes advisable to have an actual part as a working standard. This becomes particularly important when the standard part is spectrally different, as with different paint technologies or extruded plastics. An approved part as a standard should be considered for parts that are highly contoured and being compared to a flat standards panel. In all cases of color comparison, whether samples are flat or curved, painted or extruded, body panels or parts, the visual evaluation process is essential. The visual acceptance/rejection data can be correlated to multi-angle measurements and this is the preferred way of establishing pass/fail tolerances.

CONCLUSIONS

Multi-angle color measurement is an essential tool in identifying and reducing color variation in processes that use metallic, pearlescent and special effects color. There are also several options available in using the instrument to reduce the variation amongst standards, the key being to use the multi-angle spectrophotometer and a selected population of standards to quantify the actual target standard. Throughout the paint manufacturing process several conditions may warrant the development of an offset standard. This too should be a physical and quantifiable target. In many of the processes, several variables relating to processing, application and curing of color on car bodies and components need to be monitored and related to color measurements. Visual assessment information should be gathered in conjunction with multi-angle measurements so that objective color evaluation and development of meaningful tolerance can be achieved.

Success is total assembly color harmony with no color mismatches and continuous, consistent color from front to back, side to side, from top to bottom and from body to parts.

REFERENCES

1   F. W. Billmeyer, Jr., M. Saltxman, **Principles of Color Technology**. 2nd Edition, *John Wiley & Sons*, New York, NY, 1981.
2   D. B. Judd, G. Wysczecki, **Color in Business, Science and Industry**, 3rd Edition, *John Wiley & Sons*, New York, NY, 1975.
3   W. S. Stiles, G. Wysczecki, **Color Science: Concepts and Methods, Quantitative Data and Formulae**, 2nd Edition, *John Wiley & Sons*, New York, NY, 1982.
4   Timothy Mouw, Sphere versus 0/45, A Discussion of Instrument geometries and Their Areas of Application, X-Rite, Incorporated, 1994.

An Investigation of Multi-angle Spectrophotometry for Colored Polypropylene Compounds

D. Jeffery Davis
Exxon Chemical Company/Mytex Polymers

ABSTRACT

Multi-angle or goniospectrophotometers have been in existence for over fifty years. Recently, advances in instrument design, coatings technology and the growth of special effects pigments have pushed their use into the mainstream. Although they have gained prominent use in the coatings and cosmetics fields, there has been little application of these instruments for conventional colored plastics.

This paper studies multi-angle spectrophotometry as applied to integrally colored filled polyolefins for automotive applications. The paper begins by examining instrument capabilities and limitations using textured automotive parts. It also explores use of multi-angle spectrophotometry to improve the quality of automotive color matching in filled polypropylene systems.

BACKGROUND

Engineered polypropylenes have gained rapid acceptance in the automotive marketplace over the last eight years. There are several reasons for this. Polypropylene is light in weight, chemically resistant, recyclable, has favorable NVH characteristics, and lower raw material cost versus most competitive materials. Another key attribute is good weatherability and the potential for "molded in color". These combined properties can lower system costs versus competitive materials, especially those which require painting.

By 1993, there was widespread industry acceptance of engineered polypropylene, with nearly every OEM (foreign and domestic) having specifications for and using at least one grade. A great deal of effort was channeled into the development and certification of these resins, however, much of the color technology was carried over from unfilled polyolefins. In

1993, we began work towards improving the quality and consistency of our color match efforts. On several occasions we had noticed some colors matched very well on a "face" evaluation, but did not match as well under "flop" conditions. Face color can be defined by viewing the object from an angle nearly perpendicular to the sample surface. A flop evaluation can be made by rotating the sample away at the top by approximately 90 degrees. The actual angle of view for flop will be somewhere around 25 degrees. Color travel can be observed by viewing the sample while rotating from the face to the flop position. Frequently, we observed the flop color would be darker, bluer and redder than the face color. We began trying to compensate for this change by making the matches slightly lighter and yellow on face. This way, they tended to look closer on flop. We were able to measure the color using a sphere type spectrophotometer and record the positive DL* and Db* when the sample looked visually acceptable. This value would then become an offset center point for our color tolerances. The technique was successful, although it was pretty much trial and error.

In 1994, we decided to evaluate a multi-angle spectrophotometer from X-rite Incorporated, to see if the instrument would be able to measure and thus better characterize what we see visually. We felt that if we could measure color at the flop angles we could compare various submissions and formulate better matches with regards to both flop and color travel.

EXPERIMENTAL

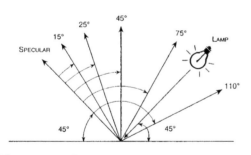

Figure 1 MA68 illumination and viewing angles.

Although both three and five angle spectrophotometers are available, we selected the X-Rite MA68 five angle instrument for our work. On this instrument, illumination is at a 45 degree angle to the sample surface. The specular component is then away from the surface at a 45 degree angle opposite illumination. The five viewing angles are 15, 25, 45, 75 and 110 degrees (above specular). The 15 and 25 degree angles are sometimes referred to as flash angles because they are near specular (the reflected light of illumination). The 75 and 110 degree angles are referred to as flop angles. Figure 1 shows the relationship of illumination and measurement for the MA68.

For this work, all samples were either injection molded automotive plaques or parts molded in accordance with our recommended processing conditions. Several different textured and smooth surfaces of varying gloss levels were evaluated. CIELAB color difference units were used exclusively. Finally, a 60 degree gloss meter was used to make sure gloss readings were comparable.

RESULTS AND DISCUSSION

Multi-angle spectrophotometers by their very design are directional instruments. Hence, we chose to begin our study by examining the effect of orientation, texture and gloss on our measurements. We chose a neutral gray textured automotive master with a low gloss level and positioned the spectrophotometer on the surface to get a measurement. We then successively rotated the spectrophotometer and took measurements at 90, 180, and 270 degrees to the original reading. We observed differences in L* value of up to 2.0 CIELAB units at the 15 degree flash angle. We also observed differences in a* value up to 0.4 units and b* value up to 0.8 at other angles. These values were observed on a grain that was not obviously directional. A highly directional grain would have had an even greater effect on the color differences observed. These differences seem significant versus normal automotive tolerances. We decided to set up a procedure to record the orientation of the instrument and surface texture for each reading. This was essentially a quick sketch of the plaque showing location(s) of measurement and orientation of the spectrophotometer. This would help with repeatability issues and allow us to go back and re-read any interesting "discoveries" at a later time.

At first we were concerned by the effect of orientation on the readings, but we realized that a similar orientation normally occurs during color reviews. Most OEM's have adopted angled overhead lighting with an angled viewing perch, as presented in the Detroit Color Council Bulletin No. 3, "Procedure for Visual Evaluation of Interior and Exterior Automotive Trim." Typically during the reviews, the top of the master plaque is positioned up and away when viewing. Many OEM's now require color submissions to be molded using a standardized plaque with the same textures and gloss level as the master. Aligning the plaques in this manner essentially fixes viewing orientation. The color will usually be checked on face, flop and for color travel. Many color reviewers also now recognize the effect of material flow direction on color, and will align the gates so that flow is in the same direction. Thus, it becomes important to match the geometric design of your plaque mold with the master, to minimize appearance differences.

When we started this work, we had hoped to address the correlation of illumination/measurement geometry for the multi-angle instrument versus the visual evaluation conditions. This was suggested to us by the fact that we had customers with different conditions for visual assessment as many molders still utilize the conventional light booth. The two systems have different illumination geometry's and intensity levels and we have seen this affect visual assessment in the past. This work would require precise measurement of the angles involved and we concluded this was beyond the scope of our work. In the end, we came to view the multi-angle unit as a tool to measure differences and help us correct those that we saw as "visually important."

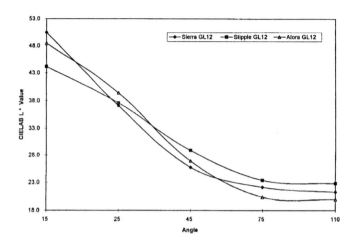

Figure 2. L* value vs. angle - textures.

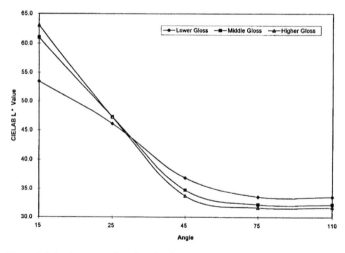

Figure 3. L* value vs. angle - gloss level.

REFLECTANCE CURVES

Reflectance curves are the foundation of successful automotive quality color matching. We thought it was only natural to begin by looking at the reflectance curves of some pigments using the multi-angle spectrophotometer. For interference pigments, the shape of the curve, and in fact, the observed color can change with the viewing angle. Fortunately, we found that with traditional pigments as used in the automotive plastics industry, the curve shape remains fairly constant. The curves do shift upwards for the flash angles and the flop angles have the lowest reflectance. We noticed some minor elongations of reflectance peaks at the flash angles, but no major changes to the curve shapes.

EFFECT OF TEXTURE

We studied the effect of texture on multi-angle readings by measuring various textures of OEM master plaques. Not surprisingly, texture which effects the way light is reflected at the surface, can have a large effect on the readings. Figure 2 shows L* value versus angle for three textures on a dark gray master plaque. In this case, the lines cross, indicating the sample that is lightest when viewed from one angle is not the lightest when viewed from another angle.

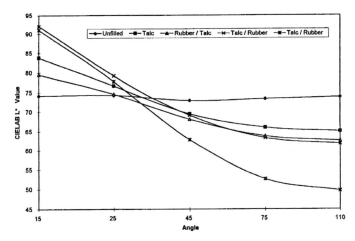

Figure 4. L* value vs. angle - filled polypropylene.

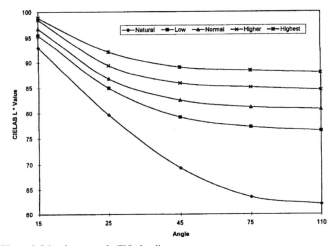

Figure 5. L* value vs. and - TiO$_2$ loading.

EFFECT OF GLOSS

Figure 3 shows L* value versus angle for three gloss levels on a basic sandblast texture. The lowest gloss level was the lightest at the 45, 75 and 110 degree angles, however, it was the darkest at 15 and 25 degrees. The highest gloss level was the darkest at the flop angles of 45, 75 and 110, but was lightest at the flash angles. The lower gloss sample was 2 - 3 units lighter than the higher gloss from 45 through 110 degrees. If the gloss levels had been further apart, the color differences would have even been larger. We found that gloss level can have a significant effect on multi-angle color readings.

EFFECT OF COLOR

We thought it would be interesting to read several arbitrary colors using the multi-angle spectrophotometer to determine if there was anything unique about how a color would behave when measured from the different angles. We selected red, blue, green, beige and gray master plaques and plotted L*, a* and b* versus angle. If you examine the a* and b* versus angle plots and think about the values, you may be able to guess what the color is, but there is nothing that makes the color recognizable as in the case of a reflectance plot.

After familiarizing ourselves with the instrument, we set out to evaluate several filled polypropylenes. Figure 4 shows L* value versus angle for several natural filled polypropylene grades. We found that for all grades, L* decreases as observation moves towards the flop angle. Note that inclusion of unfilled polypropylene is not really valid due to spectral losses (not opaque), but is shown for comparison.

After reviewing the plots of the naturals, we wondered if the filled grades might retain some of the flop even after coloring. One of the challenges in coloring the higher filled grades is to make sure you are covering the filler, without wasting pigment. You do not want the filler to act as one of the colorants in the formulation. Thus, we decided to conduct a loading study to determine the level of colorant necessary to eliminate or minimize flop for various filled grades.

Figure 5 shows the sensitivity of L* value versus angle at various TiO_2 loadings for one of our materials. Curves were also generated for a* and b*, and depending on the filler type/color, they were every bit as revealing as the L* curves. As TiO_2 loading was increased, the curve flattened out in the 45-110 degree region, showing reduced color flop. Different grades required different levels of TiO_2 and some were not reduced as much as we would have liked. For some really dark fillers, all you could do was seek to minimize the flop, or look to find a whiter filler. We decided to see what would happen with carbon black and other color blends. As expected, the carbon black colored the material dark. It did not, however, help flatten the L* versus angle curves. Further study in several grades with various colors supported the usefulness of the multi-angle spectrophotometer to help identify the optimum colorant loading for filled grades. Figure 6 shows an example where L* flattens out in the 45 to 110 region as colorant loading is increased. This was important in this case, because the OEM master was also flat in this region. As you might guess, the more work we did with colors, the more important the a* and b* versus angle curves became.

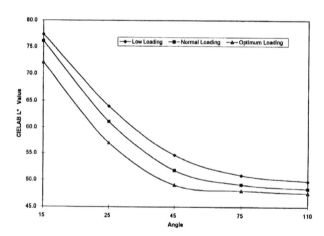

Figure 6. L* value vs. angle - colorant loading.

After concluding we could have some control over the degree of flop by adjusting the colorant loading and in some cases, by substituting pigments in the color match phase, we were anxious to apply multi-angle spectrophotometry to some real world cases. The follow-

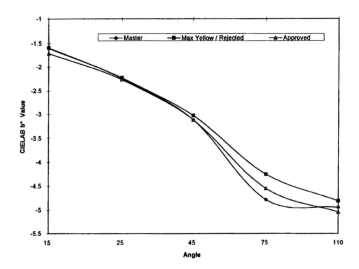

Figure 7. b* value vs. angle - OEM gray.

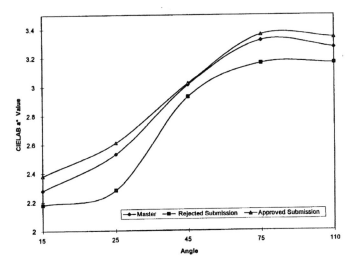

Figure 8. a* value vs. angle - OEM beige.

ing four cases are examples where the multi-angle spectrophotometer was helpful in obtaining color approvals.

Some of these cases deal with actual parts where curved surfaces allow the eye to integrate the color it sees, as it views many angles simultaneously. Flat color plaques in that regard actually represent a worst case scenario, where each angle is viewed discreetly. We also found that as with most color problems, if the hue is right, the eye will forgive minor differences in lightness/darkness.

Our first example is an OEM gray color. Figure 7 is a plot of b* value versus angle for the master, a rejected first submission and the approved submission. The rejected color was actually visually approved as maximum yellow, but was later discarded when we were able to improve the visual appearance. Notice at 45 degrees, the rejected and approved samples are almost identical, but at 75 degrees the rejected sample is about 0.5 CIELAB units yellow. The approved sample is slightly yellower than the master at 75 degrees, but goes very slightly bluer at the 110 degree flop angle. Both parts were the same grain and gloss level. Visually, the second sample looked much better and was easily approved. In this case, we were fortunate enough to have a smooth surface from which to also

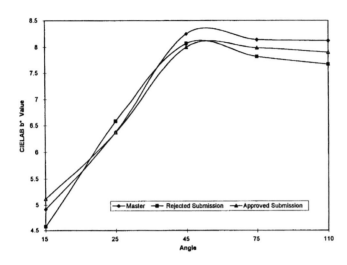

Figure 9. b* value vs. angle - OEM beige.

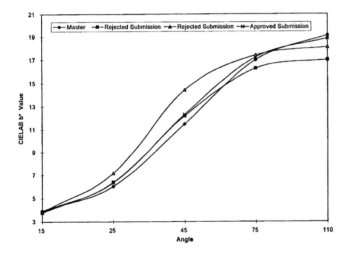

Figure 10. b* value vs. angle - OEM red.

take measurements. It is interesting to point out that we observed the same result on the smooth surface as we had for the texture part.

Our second example is an OEM beige where the initial part submission was rejected after we received visual plaque approval. Figure 8 shows a* versus angle. The part was rejected for being too dark, blue and green. Visually on face, it appeared a little green and dark, but appeared mostly blue and dark on flop. The plot clearly shows the color as 0.1 to 0.3 units greener than the master over the range of angles. The approved sample is almost exactly on target at 45 degrees and much closer at the rest of the angles. Figure 9 shows b* versus angle for the same two samples. The rejected submission is about 0.2 units blue at 45 degrees, but almost 0.5 units blue at 110 degrees. It is interesting that at the 45 degree angle, the approved sample actually measured slightly bluer than the rejected part, however, it was yellower at the flop angles of 75 and 110. Visually, the second submission did not flop as blue and was approved.

The third example is an OEM red. Figure 10 shows b* versus angle for two rejected samples and an approved sample versus the master. The two rejected samples contain different

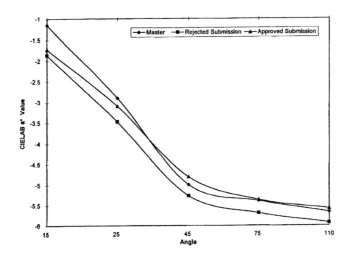

Figure 11. a* value vs. angle - OEM teal.

pigment systems and have different curve shapes from the master and approved samples. In this case, our initial attempts came up blue on flop, even though one submission was several units yellow at 45 degrees, the normal face angle. After a few submissions, we were able to get a very close visual match and gain approval on this difficult color. In fact, as the plot for the approved lot shows, the largest delta B value happened to be at 45 degrees.

Our final example is an OEM teal. Our initial part submission was rejected as too green and yellow. Figure 11 shows a* value versus angle for the master, rejected and approved submissions. The plot clearly shows the first submission is too green and after correction, the curve is much closer to the master. Notice that at the 25 angle the approval is slightly greener than the master, however, it is redder at 45 and 110. At 75 degrees, the two are almost identical. Visually the second submission was much improved.

CONCLUSIONS

Gloss, texture and measurement orientation can produce large differences in readings obtained from the multi-angle spectrophotometer. Care must be taken to make sure that samples are comparable for texture and gloss. Even though this is not always as easy as it sounds.

We found the multi-angle spectrophotometer to be very helpful in the color match process, by allowing us to quantify color differences from various viewing angles rather than relying on a purely visual evaluation. The multi-angle spectrophotometer gave us a more complete picture than sphere or 0/45 instruments alone. Plots of L*, a* and b* versus angle were helpful for illustrating color flop in both matches and trouble shooting situations. In some cases, it appeared possible to change the shape of L*, a*, or b* versus angle plot by changing the formulations. In other cases, the curve could be shifted or rotated to balance the visual appearance, but the basic curve shape did not change. In either case, the multi-angle spectrophotometer served as a tool that aided in the visual approval process. Plots of L*, a*, b* versus angle for colors do not immediately reveal the color in the manner that reflectance plots do, however plots comparing individual pigments can sometimes be helpful.

Finally, while it is doubtful that multi-angle spectrophotometers will ever replace the sphere or 0/45 units which are dominant in automotive plastics color matching, they can serve as a tool to facilitate color approvals for certain applications.

ACKNOWLEDGMENT

I wish to thank my colleagues Bill Frantz of Ametek and George Kalantzakis of Nippon Pigment, USA for their help. Special thanks also go to Brian Teunis and Bob Santine of X-Rite Incorporated for their help with the MA68 spectrophotometer. Finally, I am grateful to Exxon Chemical Company and Mytex Polymers for allowing me to present this work.

REFERENCES

1.   Harold Hunter, **The Measurement of Appearance**, 2nd Edition, *John Wiley & Sons*, New York, 1987.
2.   Procedure for Visual Evaluation of Interior & Exterior Trim, Detroit Color Council, Bulletin No. 3, 1994.
3.   Metallic & Pearlescent Effects in Automotive Coatings, X-Rite Incorporated.
4.   Color First Production Shipment, CQ1, Ford Motor Company Corporate Design Supplier Self Certification, 1994.

Color Concerns in Polymer Blends

Bruce M. Mulholland
Hoechst Celanese Corporation

ABSTRACT

Whether you are working with neat polymers or blends of polymers, color and appearance must be engineered just like any other desired thermal or mechanical property. The ability to achieve the desired color can be adversely affected by the base polymer itself or the combination with other polymers, modifiers, additives or stabilizers. Even if the color can be achieved in the blended system, other performance attributes such as UV stability, flammability, or mechanical properties may be adversely affected as well. This paper looks at some of these color concerns in coloring polymers and blends.

INTRODUCTION

To many people, color is at best a necessary evil. The coloring process amounts to adding a contaminant (the colorant) to a perfectly good polymer system to achieve a color while reducing all other properties! This perception can be transformed into reality when color is an afterthought in the whole product development cycle. That cycle typically progresses as follows:

1 the customer has a problem
2 a new resin formulation is developed to solve it
3 the new resin (in natural, uncolored form) is molded and tested
4 the new resin indeed solves the problem
5 the customer now reports they need the resin in dark green
6 the color group formulates the dark green using three times the normal pigment loading
7 the new resin in dark green is molded and tested by the customer
8 the new resin in dark green no longer solves the problem due to strength and warpage issues
9 back to step 1!

s can be avoided if the color and appearance of the resin are engi-
esired thermal and mechanical properties. That way, when the de-
properties are met, so is the color (at least the closest achievable color
cause one must also be aware that including the colorability of the resin early in
development process is no guarantee that all desired colors can be achieved. What this
work will accomplish is to help define problems and limitations in achieving colors up front.
And if there are limitations, this information can then be used with the customer to help select
colors that are achievable without detrimental side effects.

None of this of course is new information. But it can be forgotten. The purpose of this paper is to present information that unequivocally shows that color should not be an afterthought in the development cycle. Discussion will be focused on describing adverse effects
on colorability attributed to the base polymer itself, other blended polymers, modifiers, additives and stabilizers. [Note: since this paper is written by a colorist, information will be presented in the spirit of the pure colorant being adulterated by the polymer system and all of its
additives!]

VISUAL COLOR PERCEPTION PROCESS

Before we can discuss color concerns in polymers and blends, it is important that the reader
has an understanding of the visual color perception process. In order to see color, three things
must exist: a light source, an object and an observer. For purposes of this paper, describing the
light source and observer are not included other than to say we assume the light source contains all wavelengths (i.e., white light). The object in our case is an article molded from our
colored resin. A beam of light reaches the surface of our object. A portion of the light is reflected due to the surface interface and is called the specular reflection or gloss component.
The remainder of the light penetrates the surface of the object where it is modified through selective absorption, reflection, and scattering by the colorants, polymers and additives. Selective absorption and reflection by wavelength create color. For example, if an object absorbs
all wavelengths of light other than blue, blue light will be reflected or transmitted and the object will appear blue.

Scattering occurs when the light beam contacts particles or regions with refractive indexes different from that of the polymer. If scattering occurs equally at all wavelengths with
no absorption, the object will look white. The amount of scattering depends on the magnitude
of difference in refractive index between the polymer and additive and the particle size of the
scattering constituent. These relationships are shown in Figure 1 and Figure 2, respectively.[1]

It is important to understand the mechanism of scattering because the majority of the
problems in coloring polymers and blends can be related to the intrinsic whiteness or scattering of the polymer system itself. Increasing the amount of scattering in our resin system will

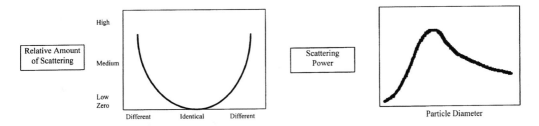

Figure 1. Scattering as a function of refractive index difference.

Figure 2. Scattering as a function of constituent particle size.

increase the amount of diffuse reflection (white light) which is mixed with the reflected colored light generated by our pigment interactions. This mixing will dilute the color strength and the color of our object will appear lighter and less bright to the observer. An analogy can be percolated from everyday life: coffee. Coffee without cream is said to be black (actually more like dark brown) due to light absorption by the coffee extract dispersed in water. Adding cream to the coffee increases the light scattering and the color is transformed from dark brown to light tan. And it is obvious that once the cream is added, it is impossible to color the mix with more coffee extract to achieve the original black color.

So is the case in polymers and blends. As will be shown, some polymer systems can impart so much light scattering that certain colors can no longer be achieved. Or if they can be achieved, other properties may be adversely affected such as impact strength and cost. In either case, the practical color gamut or palette that is obtainable with this particular resin system is reduced. The discussion below presents the effects that the polymer and its additives can have on colorability.

POLYMER TYPES

TRANSPARENT RESINS

From the discussion above, it should be intuitive that transparent resins such as polystyrene or polycarbonate do not scatter light and therefore can achieve the most brilliant colors. Colors are generated through transmitted or reflected light without having to mix with white light from diffuse reflection since none exists from the polymer. The difficulty here is making the system opaque (no light transmission). If opacity is required, a scattering opacifier such as titanium dioxide is needed and apparent color strength will be reduced.

TRANSLUCENT & OPAQUE RESINS

Most polymers fall in the class of translucent resins. These include acetal, polyamide, polybutyleneterephthalate (PBT), polyethylene and polypropylene as examples. There are very few neat polymers that are truly opaque (this depends on thickness as well). Liquid crystal polymer (LCP) is an example of a typically opaque polymer. It is theorized that these semi-crystalline and crystalline resins will scatter some portion of incident light due to spherulitic crystal structure and the amorphous-crystalline region interfaces themselves.

Table 1. Contrast ratios of selected resins

Resin	L-value* white	L-value black	Contrast ratio @ 3.2 mm
Polyamide 6,6	85.65	66.75	0.78
Acetal	93.01	82.05	0.88
PBT	92.70	87.14	0.94
PBT (annealed)	92.80	90.94	0.98
LCP	84.10	84.10	1.00

*specular excluded, expressed in HunterLab units

The degree of translucency can be measured by calculating a contrast ratio. This number is the ratio of the L-value in color measurement obtained when backing the natural sample with a black tile divided by the L-value of the sample backed with a white tile. A completely opaque sample will have a contrast ratio of 1.00. Contrast ratio measurements are specific to the thickness of the sample and the intensity of the light source. Contrast ratios for several resins are presented in Table 1.

The data in Table 1 shows that for these crystalline resins, the contrast ratio generally increases with increased degree of crystallinity. Polyamide 6,6 has the lowest contrast ratio (less opaque) compared to liquid crystal polymer that has an extremely high crystallinity and is completely opaque. Furthermore, the two PBT samples show that the annealed sample (higher crystallinity) is also more opaque. The higher opacity indicates that more light scattering is occurring by the polymer and typically a more restricted color gamut will be achieved. Polymers with a high degree of intrinsic whiteness will exclude deep, dark colors and bright, high chroma colors from the achievable color gamut.

To illustrate this effect on colors, three single pigment colors (blue, red and yellow) were developed in three different resin systems. Blue and yellow were produced in acetal (POM), polyphenylene sulfide (PPS) and LCP. The red color was prepared in polyamide 6,6, PPS and LCP. The acetal and polyamide resins are translucent while the PPS and LCP are opaque at the 3.2 mm sample thickness. In all three colors, the more translucent resins produced visually more brilliant, higher chroma colors than the more opaque resins with increased diffuse scattering.

Table 2 presents color data generated from these same samples. Numerically, PPS and LCP produced lighter colors (higher L* values) and/or lower chroma (C* values) compared to the acetal and polyamide controls which is consistent with what is seen visually. The base resin light scattering will increase the L* value and decrease chroma in these colors. What's more, the PPS and LCP were evaluated from 2 times to 15 times the pigment loading (relative concentration) of the acetal and polyamide, and this still holds true. Another indication of the increased scattering from the PPS and LCP is the relative strength data of Table 2. As shown, even at the higher relative pigment concentrations, these resins produce colors that appear only 16 to 50% as strong.

Table 2. Relative Color Strength in Various Resins

Color	Resin	Relative concentration	L*	C*	Relative strength, %
Blue	POM	X	37.05	55.01	
	PPS	3X	47.78	34.34	42
	LCP	3X	61.60	27.93	16
	LCP	6X	56.14	35.09	26
Red	PA	X	35.91	41.61	
	PPS	3X	44.43	33.67	48
	LCP	3X	56.87	34.13	23
Yellow	POM	X	77.52	77.94	
	PPS	2X	69.35	52.45	50
	LCP	15X	78.13	51.03	31

This discussion illustrates the effect light scattering of the polymer has on the resulting color. In real life, if LCP is required for a particular application, the examples in Table 2 show that deep, dark colors or bright, high chroma ones cannot be achieved. If the customer is currently using these higher chroma colors in other resins such as polycarbonate, he needs to be educated that these same colors can not be achieved in LCP.

The other limiting factor in coloring a specific polymer is the stability and suitability of the colorant for a particular polymer and application. A general rule of thumb is that as the recommended processing temperature for a resin increases, the number of colorants that can withstand those temperatures decreases. Most of the colorants that drop out are organic pigments that are typically used to achieve the bright, high chroma colors or the deep, dark colors. Furthermore, polymers with harsh chemical environments like PVC, acetal or polyamide limit colorants based on chemical stability. Conversely, the chemistry of the colorant can also render the polymer unstable, making it unsuitable for use as well. Finally, end use requirements such as agency compliance (FDA) or UV stability will further restrict the number of colorants that can be used in a specific polymer. Thus the achievable color gamut for a resin system not only depends on the light scattering of the polymer but also on the types of colorants that can be utilized.

POLYMER BLENDS

One can assume that blends of polymers will be more difficult to color than any component by themselves. Diffuse reflection can increase due to internal light reflection or scattering at phase interfaces if the polymers are at least partially immiscible or their refractive indexes are significantly different. Blends of translucent polymers are typically more opaque than either resin alone. Furthermore, colorant stability (thermal or chemical) can be adversely affected by the presence of the other polymer(s). As in the case of neat polymers, both circumstances will result in a restricted achievable color gamut for the polymer blend. An example of a prominent polymer blend is GE's Noryl® (PS/PPO) which certainly colors much differently than the polystyrene component by itself.

Blends of polymers can pose their own unique problems as well. An example is cases where colorants exhibit preferential dispersion to one of the polymer phases. The other polymer phase remains virtually uncolored. Macroscopically, this may not be a problem as the molded part appears uniformly colored. But even at this level, if wall thickness is very thin, color striations may become apparent. Other performance measures may be adversely affected as well. At the microscopic level, since all of the colorant is dispersed in one phase, impact strength and other properties may be reduced at pigment concentrations that are much lower than expected. This would primarily occur in blends where the colorant prefers the resin phase that provides the toughness to the blend.

POLYMER ADDITIVES

A wide variety of additives and modifiers are incorporated into polymers and polymer blends to tailor specific properties. Unfortunately, these additives can also impact the colorability to the total resin system. Listed below are some common additives and modifiers used in polymers and a short discussion of their typical effect on colorability.

ANTIOXIDANTS

In general, antioxidants will have little effect on colorability since they are typically used at low levels. At higher levels, they may increase light scattering and impact colorability depending on the polymer type. There are remote instances where antioxidants have been linked to problems with graying bright colors or "pinking" whites. But these are very polymer specific and usually result from a chemical instability within the system.

ANTI-STATS

These additives are designed to be present on the surface of the molded part to achieve the full anti-static benefit. Furthermore, they are typically used at higher levels than other additives such as antioxidants. Therefore, anti-stats are likely to increase light scattering making it more difficult to achieve the higher chroma colors.

COUPLING AGENTS

Coupling agents such as silanes and titanates will increase light scattering. Both types can impact colorability if they are incorporated at high levels.

FLAME RETARDANTS

Table 3. Effect of flame retardants on colorability (fr resin versus neat polymer)

Color	DL*	DC*	Relative strength, %
Bright yellow	2.3	-3.5	56
Bright red	4.9	-7.1	53
Medium tan	7.0	-2.4	57
Bright orange	4.0	-1.5	74
Royal blue	6.3	-3.2	78
Dark green	9.3	n/a	52
Medium gray	5.0	-0.5	n/a
Black	3.4	-0.9	72

Typical flame retardant formulations will include an antimony compound, a bromine compound, and possibly a drip suppressant. All three additives will significantly increase light scattering and reduce the color gamut of the resin. Table 3 presents color data to show this effect. Eight flame retardant polyester colors are listed with lightness, chroma and strength differences calculated versus the same colors in neat polyester resin. All colors are lighter (positive DL*) and duller with lower chroma (negative DC*) with the presence of the flame retardant system. Relative color strength data also shows that the colors are weaker and less intense. It would be difficult if not impossible to exactly match the neat polymer color in the flame retardant system due to the light scattering of the additives.

FOAMING AGENTS

The author has little experience with foamed resins. However, one would speculate based on all of the above information that foaming would increase light scattering since the molded parts would contain a complex cell structure with all of its polymer/gas interfaces.

HEAT STABILIZERS

Antioxidants can be referred to as heat stabilizers and were previously discussed. Other heat stabilizers are metal complexes used in PVC and polyamide. The largest impact on colorability is with the metal complexes used to increase the continuous use temperature rating in polyamide. In that resin, these metal complexes can significantly reduce colorability due to discoloration *via* reaction with the colorant, or by thermal degradation. Light colors in heat stabilized polyamide are virtually impossible to control since a slight change in residence time or temperature will significantly drive the color to tan or brown. The presence of oxygen will accelerate this. Darker colors are less affected and are preferred in heat stabilized polyamide.

IMPACT MODIFIERS

Table 4. Effect of impact modifier on colorability

Color	DL*	DC*
Gray	4.5	-1.8
Bright red	2.8	-2.4
Bright yellow	3.6	-4.0

This class of additives covers a broad range from butadiene to acrylic polymers. Since these additives are polymeric in nature, diffuse reflection will occur at the polymer/modifier interfaces similar to polymer blends. Again, this will result in colors that appear lighter and duller. Table 4 contains three examples of impact modified colors compared to the neat resin without modifier. As expected, the impact modified colors are lighter and have lower chroma. In practice, ABS or HIPS would have a more restricted color gamut compared to their transparent SAN and PS base polymers.

LUBRICANTS/MOLD RELEASES

Table 5. Effect of impact modifier on colorability

Color	DL*	Relative strength, %
Bright red	1.6	85
Dark blue	2.7	80

Like anti-stats, lubricants and mold releases are designed to reside on the molded part surface. Therefore, light scattering can be increased depending on the chemistry and concentration of additive. Generally this effect is not a problem unless these additives are used at high levels, or they have significantly different refractive indexes from the base polymer. Table 5 shows color difference data for two colors containing a high level of lubricant versus a polymer with no lubricant. Increased light scattering is evident by the lighter color (positive DL*) and lower relative color strength.

REINFORCING AGENTS

Reinforcing agents are typically either a mineral such as talc and calcium carbonate, or fiberglass. As expected, mineral types used at their high levels can scatter a large portion of the incident light depending on the refractive index and particle size. In mineral filled resins, it is usually impossible to achieve deep, dark colors such as chocolate brown or forest green.

Table 6. Effect of fiberglass on colorability

Color	DL*	DC*
Gray	-1.8	-0.2
Dark blue	-1.5	-0.7
Bright yellow	-5.9	-8.2
Dark green	-3.5	-2.9
Bright red	-1.7	-0.6

Bright colors are also made duller. Because a high amount of mineral is generally used, bright white colors such as appliance white can also be difficult to achieve.

Fiberglass poses a different problem in that colors are typically darker and more dingy looking in glass reinforced resins versus their unfilled counterparts. Table 6 lists color difference data for several 30% glass reinforced PBT colors versus the same color in unfilled PBT. The darker, dingier look is evident by the negative lightness and chroma difference values. Brighter, high chroma colors in glass reinforced resins either require significantly more colorant to achieve (at higher cost) or can not be exactly matched due to the resin. Overcoming the darkness makes bright white colors difficult to achieve as well. Contributing factors to the dark, dingy appearance are most likely the glass sizing agent and the elevated processing temperatures for glass reinforced resin versus the unfilled product.

Fiberglass adds another problem to coloring: mechanical property retention. A number of widely used colorants will abrade glass fiber length or can affect fiber wetting significantly reducing properties. For example, in 30% glass reinforced PBT, certain colorants can reduce tensile strength up to 20% and notched izod impact by as much as 30%. And this reduction occurs using colorant concentrations that would be termed typical and not excessive. Mechanical property retention may further limit the types and numbers of colorants that can be used in the glass reinforced resin, further limiting its achievable color gamut.

UV STABILIZERS

A large number of colored polymers and polymer blends are used in applications where UV stability is important. First and foremost, many common colorants do not possess satisfactory lightfastness to be used in UV applications. Therefore, the color gamut can be reduced due to limited availability of colorants with acceptable lightfastness in certain polymer systems. UV stabilizers themselves can also impact colorability. Many impart a yellowish tint to the polymer that must be overcome. Also, some do increase light scattering making the more popular automotive maroon and dark blue colors difficult to achieve.

CONCLUSIONS

Hopefully, the discussion and examples presented above have clearly shown that color should be engineered along with other desired thermal and mechanical properties. That way, either the desired color can be formulated to minimize property loss, or the closest achievable color can be developed. And understanding the color limitations early on can help steer customers to those colors that can be achieved, which may be different from those actually desired.

Most of the discussion has been centered around just the neat polymer or the polymer plus a modifier or additive. Coloring issues further escalate when polymer blends are used or when multiple additives are incorporated. Would it be difficult, and would the color gamut be somewhat restricted, if one was to color an impact modified, flame retardant, glass reinforced PBT/polycarbonate alloy? The answer now better be a resounding YES!

REFERENCES

1   F. W. Billmeyer, Jr. and M. Saltzman, **Principles of Color Technology**, 2nd Ed., *John Wiley & Sons*, New York, p. 12 (1981).

The Effect of Pigments on the Crystallization and Properties of Polypropylene

S. Kenig, A. Silberman, I. Dolgopolsky
Israel Plastics and Rubber Center, Technion City, Haifa 32000, Israel

ABSTRACT

The effect of pigments having different chemical composition on the crystallization process of polypropylene, PP, was studied by means of differential scanning calorimeter in isothermal mode. Analysis of the experimental results was carried out using Avrami's equation to obtain the crystallization parameters. Furthermore, Lipatov's approach was applied for evaluation of the thickness and content of the transition layers formed as a result of the interaction between the PP and the various pigments. Results have shown that depending on their composition the pigments act like nucleating agents affecting the degree of crystallinity, the structure of the amorphous phase and the interphase between the PP and the different pigments. The resultant morphology has been associated with the mechanical properties of the various pigmented PP compositions.

INTRODUCTION

The use of pigments in polymer compounds has become widespread in the plastics industry for aesthetic as well as for functional purposes. Polypropylene (PP), being semicrystalline polymer, is highly sensitive to incorporation of additives.

Pigments used to color polypropylene, as used also in other polymers, are organic or inorganic compounds. They are usually employed as concentrated master batches dispersed in a carrier resin. The carrier resin may be based on polypropylene or another polymer. The final pigment concentration, in the end product is in the range of 0.1-2.0% depending on the color sought.

As pigments affect the polymer properties, a number of investigations have dealt with this subject, among them the papers of Lipatov[1,2] and others. They studied the effect of small concentrations of fillers on the change in thermal and mechanical properties of polymer compositions. This effect was noted also for pigments.

The nucleation effect of some pigments in the crystallization process was investigated in references.[3,4] Moreover, Jacoby and Bersted[5] found that certain pigments serve as nucleators for PP, affecting the crystallization process, structure and crystalline form. The pigments influence on the crystallization process in PP filaments was studied by Lin and Shou.[6] They noted that the nucleating effect of different pigments was manifested by a crystallization temperature increase.

The present study attempts to relate the pigment chemical composition with the structure of pigmented polymer and its resultant mechanical properties.

MATERIALS AND METHODS

MATERIALS

Commercial isotactic polypropylene, Moplen F30G (Himont Corp.) was used throughout the investigation. It is a homopolymer with MFI 12 (230°C/2.16 Kg).

The organic pigments studied were:
(1) Irgazin Yellow 2GLT-tetrachloroisoindolinone (Pigment Yellow 109, No. 56284 by Ciba-Geigy);
(2) Hostaperm Red E3B-2,9-dimethylquinacridone (Pigment Red 122, No. 73915 by Hoechst);
(3) Chromophtal Green GFN-Cu-phthalocyanine (Pigment Green 7, No. 74260 by Ciba-Geigy).

PP specimens with various pigment contents were prepared as follows. In the first stage a pigmented master batch was prepared using PP pellets mixed with dry pigment (10% concentration) in a Brabender.

In a second stage the master-batch and PP pellets were molded using an injection molding machine. The final pigment concentrations in the molded specimens were 0.1, 0.5 or 1.0%, respectively.

CHARACTERIZATION METHODS

Crystallization properties were determined using a differential scanning calorimeter (DSC, Du Pont Thermal Analyser 2000). Isothermal crystallization was carried out by heating the sample up to 200°C at a rate of 125°C/min, where they were held for 10 min to remove the previous thermal history. The specimens were then quenched to the appropriate isothermal crystallization temperature.

The relative degree of crystallinity, X_t, was determined from the partial area under the DSC exotherm defined as follows:

$$X_t(t) = \int\limits^t (\Delta H / \Delta t)dt / \int\limits^\infty (\Delta H / \Delta t)dt \qquad [1]$$

where $\Delta H / \Delta t$ is the rate of heat evolution and t is the crystallization time.

Crystallization kinetics was studied using the modified Avrami equation.

$$Q_a = exp\ (-Kt)^n \qquad [2]$$

where Q_a is the fraction of the uncrystallized polymer, K is the kinetic constant and n is the Avrami exponent related to the crystal geometry and the mechanism of crystallization. Hence, taking the double logarithm of Eq. [2] gives:

$$ln[ln(-Q_a)] = n\ ln\ K + n\ ln\ t \qquad [3]$$

Thermal properties were obtained from a DSC non-isothermal (dynamic) scanning at a rate of $10°C/min$. The glass transition temperature, T_g, was obtained by scanning at a rate of $20°C/min$. The specific heat capacity, C_p, and ΔC_p value (C_p differential) was calculated from the height of the C_p curve in the glass transition temperature region, taking the tangent of the enthalpy, ΔH, vs. temperature in the isothermal crystallization curve.

RESULTS AND DISCUSSION

The effect of pigments on the behavior of PP was studied by thermal analysis in isothermal and non-isothermal modes. The crystallization enthalpy, ΔH, was measured for every isothermal run.

The temperature where the enthalpy rises sharply is a characteristic property of semicrystalline polymers. This temperature is defined as the equilibrium temperature, T_{eq}, since it characterizes the structure closest to the equilibrium state in a semicrystalline polymer.

During isothermal crystallization the actual crystallization temperature (for example $122.2°C$) was not exactly equal to the set temperature ($125°C$). This is due to the fact that the lamellar size can have only defined values, which depend not only on the conditions of crystallization, but also on the macromolecule structure.

The activation energy[7] of the crystallization process U_{act} was calculated using the half-time of the crystallization ($\tau_{1/2}$) according to the following equation:

$$lg\tau_{1/2} = lg\tau_o + U_{act} / RT \qquad [4]$$

where R is the gas constant and τ_o the material constant.

The enthalpy of crystallization, ΔH, equilibrium temperature, T_{eq}, and activation energy of crystallization, U_{act}, values are summarized in Table 1.

Table 1. Thermal properties of polypropylene

Color	Pigment content, %	T_{eq}, °C	ΔH, J/g	C_p, J/g °C	U_{act}, kJ/mol
Yellow	0.1	124.6	74.9	4.79	94.6
	0.5	125.2	77.0	3.63	100.2
	1.0	125.2	77.8	5.13	100.2
Red	0.1	132.3	82.4	7.11	71.1
	0.5	133.4	71.2	4.78	42.7
	1.0α	132.8	71.0	7.30	
	1.0β	135.6	74.6	21.0	46.0
Green	0.1	132.3	62.4	9.75	36.1
	0.5	133.4	71.2	3.80	43.6
	1.0	135.6	74.6	2.30	31.4
Neat	-	121.0	51.9	2.16	87.5

Lipatov[2] studied the interaction at the interphase between polymers and fillers. Accordingly, the fraction of molecules, in the amorphous phase absorbed on the solid filler surface, v, can be determined from the following equation:

$$v = 1 - \Delta C_p^f / \Delta C_p \qquad [5]$$

where ΔC_p^f and ΔC_p are specific heats for filled and unfilled polymers, respectively.

An extremely high value of v was obtained in the case of 0.1% content (Figure 1) which may manifest the "effect of small concentrations". Moreover the thickness of the polymer transition layer, Δr, absorbed on a pigment particle can be approximately evaluated as follows:[9]

$$[(\Delta r + r) / r]^3 - 1 = v\Phi / (1 - \Phi) \qquad [7]$$

where r is the average radius of pigment particles and Φ is the volume content of filler. The Δr values vs. pigment concentration is shown in Figure 2.

The degree of crystallinity for all PP samples were obtained from the crystallization enthalpy, ΔH, using dynamic DSC measurements, as follows:

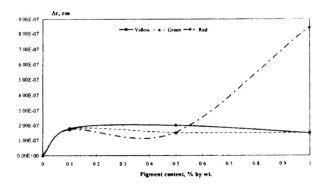

Figure 1. Thickness of polymer transition layer.

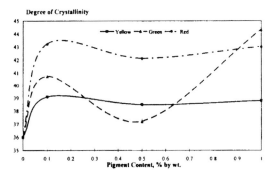

Figure 2. The effect of pigment content on the degree of crystallinity of pigmented polypropylene.

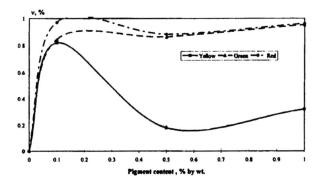

Figure 3. The effect of pigment content on the yield strength of pigmented polypropylene.

$$\phi = \Delta H / \Delta H_{cr}$$

where $\Delta H_{cr} = 163$ J/g.[8]

As can be seen from Figure 3 a substantial increase of crystallinity takes place at small concentrations for all pigments. It follows from Lipatov's investigations[1,2] that such "effect of small concentrations" exists only for active fillers. Thus it may be concluded that all three pigments studied could be categorized as active fillers.

As pointed out by various authors,[8,9] addition of pigments to plastics usually only slightly affects their strength and structure. The results presented in Figure 3 and Table 2 indicate that the above conclusions cannot be related to PP. The influence of each pigment on PP's degree of crystallinity is a specific effect, as can be concluded from the kinetics of crystallization.

According to reference[10] the primary and secondary crystallization may be defined through the experimental techniques used, and consequently the kinetic parameters and reaction order for each process may be determined. Primary crystallization is associated with the growth of spherulites, while secondary crystallization is associated with additional crystallization within the boundaries of a spherulite as well as the crystallization of material trapped between spherulites. The re-

Table 2. Mechanical properties of pigmented polypropylene

Color	Content %	Young's modulus, MPa	Izod impact, J/m	Yield stress, MPa
Yellow	0.1	1810	2.84	36.9
	0.5	1730	4.15	37.2
	1.0	1700	3.23	37.8
Red	0.1	1700	3.94	35.9
	0.5	1610	3.88	35.9
	1.0	1890	3.91	36.9
Green	0.1	1710	4.00	37.1
	0.5	2040	3.00	39.1
	1.0	1890	2.80	36.8
Neat	-	1420	2.93	35.6

action order (n_{prim}) for the first stage of crystallization-nucleation exhibits a wide distribution. Moreover, the n_{prim} values are greater than 8 and thus the process cannot be described in the frame-work of Johnson-Mehl-Avrami approach.[10] This may be attributed to the presence of a great number of small particles with high surface energy (organic pigments) which can be nucleating agents for polypropylene. Hence, the secondary crystallization process will be discussed.

The reaction order for the secondary crystallization was calculated according to Eq. [2] and is presented in Table 3. It can be observed that for yellow and green pigments as well as for non-colored PP, the n_{sec} value is constant up to temperatures 125, 132 and 120°C respectively. It follows from the Schultz approach[10] that up to these temperatures the diffusion growth control of crystallization takes place, above them interface growth is the decisive control factor.

According to Landau's phase transformation theory,[11] a sharp change of all structural parameter materials takes place in vicinity of phase transformation.

The activation energy of crystallization, U_{act}, vs. pigment content is given in Table 1. The yellow pigment addition results in increased activation energy and thus hinders chain packaging. Distinctively, red and especially green pigments enhance this process. This is due to the specific chemical structure of pigments. Isoindolinone molecules have lower polarity as a result of the symmetry of the halogen atoms, which leads to steric hindrance for crystallization of PP. The groups strongly attach to PP molecules and thus inhibit PP chain packaging. Further adsorption of PP on pigment is limited.

The symmetry of nitrogen in the green phthalocyanine pigment molecule, with an absence of complex structures, results in the dynamic (physical) adsorption of PP on the pigment surface.

The highly polar quinacridone red molecules possess donor (keto-) as well as acceptor (amino-) chemical groups. Thus chemical as well as physical absorption can be realized on these pigment surfaces. Accordingly, it may be assumed that at small concentrations the

Table 3. The reaction order of secondary crystallization in isothermal conditions

Content %	Color					
	Yellow		Red		Green	
	T_{cr}	N_{sec}	T_{cr}	N_{sec}	T_{cr}	N_{sec}
0.1	122.0	2.11	127.5	2.44	128.5	2.61
	123.7	2.82	129.5	3.02	130.4	3.00
	125.0	2.52	131.2	3.18	132.0	2.29
	126.7	3.31	132.8	3.92	134.0	5.57
	128.4	3.76	134.6	4.53	135.8	5.28
	130.0	3.35	136.5	5.72	137.5	4.45
	132.0	5.06	138.3	4.22	-	-
0.5	123.5	2.07	128.8	2.39	129.5	2.25
	125.5	2.79	131.1	2.79	131.3	2.69
	126.8	3.43	132.6	3.88	132.4	2.93
	128.5	4.93	134.4	3.78	134.1	3.70
	130.2	5.91	136.2	4.13	135.9	4.46
	131.2	5.77	137.7	4.05	138.1	7.60
1.0	123.8	2.34	128.0	2.29	129.3	2.53
	125.5	2.06	130.7	3.05	130.0	2.45
	127.3	3.85	132.5	3.40	133.0	2.14
	128.7	4.82	134.0	3.12	134.8	3.85
	130.4	5.00	135.0	3.93	136.4	5.06
	131.4	5.48	137.8	4.30	138.3	5.70
neat	116.0	1.59				
	119.0	1.83				
	120.5	1.69				
	122.0	3.18				
	124.0	3.99				
	128.6	3.52				

chemical type of adsorption predominates, while as the concentration increases it changes to physical type adsorption. All above effects may be the reason for the strong nucleation effect of red quinacridone pigment. High adsorption activity of red and green pigments is exhibited close to T_{eq}, which is higher than that of the yellow pigment. The high adsorption affinity of red is displayed in the curve (Figure 1) for 0.1% pigment content. This presents an additional evidence for its strong nucleation activity, described in references.[5,12] According to Jacoby and Bersted,[5] isotactic polypropylene is capable of crystallizing in α-(monoclynic) and β-(pseudohexagonal) crystalline forms. Higher concentrations of red quinacridone leads to β-crystal formation, characterized by lower T_{eq} and ΔH (Table 1). It corresponds with reference,[5] where the different levels of β-crystallinity were observed.

In the course of the present investigations, three main mechanical parameters were evaluated: Young's modulus, E, yield strength, σ_{yield}, and notched impact strength, a. It is known that for the semicrystalline polymers the yield value can be correlated with the degree of crystallinity. Comparison of Figure 3 and Table 4 indicates good correlation between σ_{yield} and degree of crystallinity for the yellow pigment. For the green pigment this correlation is poor, while it does not exist for the red pigment. This may be due to the fact that yielding takes place predominantly in amorphous phase.

Table 4. The effect of pigment's addition on the main mechanical properties of polypropylene

	Pigment color		
	Yellow	Red	Green
Crystalline content	0.38	0.42	0.37
Amorphous content	0.62	0.58	0.63
Amorphous phase (v and r) cm	$0.18/2 \times 10^{-7}$	$0.9/1.8 \times 10^{-7}$	$0.87/1.8 \times 10^{-7}$
Interphase type	average	perfect	defect
Modulus, MPa	1730	2040	2040
Strength, MPa	37.2	39.1	39.1
Impact, J/m	3.2	3.0	3.0

As far as the impact strength of the semicrystalline polymers is concerned, it is dominated by the amorphous phase structure. The correlation between the impact, a, and pigment content is presented in Table 4. For the yellow pigment the poor transition layer (Figures 1 and 2) corresponds with high impact values. The same correlation, but more pronounced, can be noted for the green pigment. For the red pigment on the v and Δr values exhibit dependence with the pigment content due to its strong nucleation activity. It is interesting to note that for the green pigment the smooth change of v and Δr vs. pigment content dependence is accompanied by a significant (1.5 times) decrease of impact. This may be attributed to the interphase structure between the amorphous and crystalline phases.

The adsorption of the polymer on the filler particles has the most pronounced effect on the amorphous phase. The transition layer, v, and the thickness of this layer, Δr, vs. pigment content dependencies are shown in Figures 1 and 2. The v-Δr relationships for the red and green pigments are similar, while it is distinctively different for the yellow pigment. The transition layer for the yellow samples is thicker while the v value is the same for the green and red pigments. This manifests itself in the previously discussed relationship between the pigment chemical structure and the PP adsorption on the pigments' particles.

The analysis of the effect of pigments on PP mechanical properties should refer to the crystalline phase (number and size of lamella), as well as to the amorphous phase (content and structure characteristics) and the interphase between them. As can be observed from Table 4, the yellow pigment addition leads to an increase of amorphous phase content, in comparison with other pigments, especially the red. Crystallization kinetics determines whether diffusion or interfacial control has a more complicated effect. It leads to higher interphase defects in the case of the yellow pigment compared with the red one, but lower in the case of the green pigment. The integrity of the interphase results in high impact strength accompanied by nominal Young' s modulus and yielding strength values in the case of the yellow pigment.

The nucleating effect of the red pigment is the most pronounced leading to the highest crystallinity. Diffusion crystal growth control leads to non-defective interphase. Such a morphology results in the highest impact strength and the lowest Young' s modulus and yielding strength values.

The red pigment affects the secondary crystallization while the green is a nucleating agent for the primary crystallization. Its addition leads to a high content of the strongly ordered amorphous phase and to low crystallinity, respectively. Interphase growth control leads to the extremely defective interphase. Such structure results in the lowest values of impact strength and in the highest values of Young' s modulus and yielding strength.

REFERENCES

1   Y. Lipatov, **Interface Phenomena in Polymers**, *Naukova Dumka*, Kiev (1972).
2   Y. Lipatov, **Physical Chemistry of Filled Polymers**, *Chimia*, Moskow (1977).
3   E. Hermann, *Plastverabeiter*, *18* (12), 897 (1967).
4.   G. Kaufmann, **Einfarben von Kunststoffen**, *VDI-Verlag*, Dusseldorf, 161, 285 (1975).
5   P. Jacoby, B. H. Bersted, *J. Polym. Sci.: Part B, Polym. Phys.*, **2**, 461 (1986).
6   Y. Lin, L. Shou, ANTEC/91, 1950 (1991).
7   D. W. van Krevelen, *Chimia,* **32**, 279 (1978).
8   H. Schonborn, J. P. Luongo, *Macromolecules*, **2**, 64 (1969).
9   M. Ahmed, **Polypropylene Fibers - Science and Technology**, *Elsevier*, Amsterdam (1982).
10   J. Schultz, **Polymer Materials Science**, *Prentice Hall*, Englewood Cliffs, New Jersey (1974).
11   L. Landau, E. Livshitz, **Statistical Physics** (Theoretical Physics, Part 5), *Nauka*, Moskow (1987).
12   A. Lustiger, C. N. Marzinsky, R. R. Mueller, H. D. Wagner, ANTEC'93, 2559 (1993).

The Effect of Nucleating Agents on the Morphology and Crystallization Behavior of Polypropylene

E. Harkin-Jones, W. R. Murphy
Department of Chemical Engineering, The Queen's University of Belfast, Belfast BT9 5AG
N. Macauley
Smith & Nephew Group Research Centre, York YO1 5DF, N Ireland

ABSTRACT

This paper describes the effect of nucleating agents on the morphology and crystallization behavior of polypropylene. Nucleated and un-nucleated grades of polypropylene are compared, and the ability of both titanium dioxide pigment particles and recycled material to nucleate polypropylene is examined. Nucleating agents are found to have a significant effect on crystallization behavior and titanium dioxide pigment is found to be a relatively efficient nucleating agent at addition levels as low as 0.5%. The consequences of these results for obtaining end product consistency in polypropylene are discussed briefly.

INTRODUCTION

Polypropylene has a crystal growth rate some one hundred times lower than that of high density polyethylene for example. It therefore responds well to heterogeneous nucleation by well dispersed additives present in the crystallizing melt.

Nucleating agents provide numerous growth nuclei in the melt resulting in the onset of crystallization at higher temperatures and in the formation of a fine morphology dominated by α-spherulites.

The aim of this investigation is to examine the effect of nucleating agents on factors such as crystallization onset temperature and spherulite size. The nucleating potential of titanium dioxide based pigment and that of regranulated process waste is also examined since a knowledge of such effects is essential if process and product consistency is to be achieved.

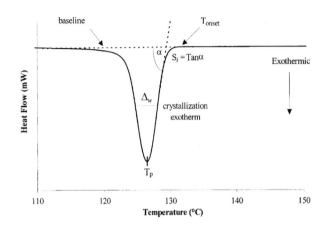

Figure 1. Illustration of typical crystallization exotherm obtained during controlled cooling of a BASF 1184L (nucleated) polypropylene sample.

Differential scanning calorimetry is used in this study to examine the crystallization behavior of polypropylene. Crystallization exotherms are analyzed according to the procedure outlined by Gupta *et al.*[1] where the parameters shown in Figure 1 are defined as follows: (i) T_p is the peak exotherm temperature; (ii) S_i is the slope of the initial portion of the crystallization exotherm which is proportional to the rate of nucleation; (iii) Δ_w is the width of the exotherm at half-height of the peak and is related to the spherulite size distribution (the narrower the distribution, the smaller is Δ_w); (iv) A/m relates to the area under the exotherm per unit weight of the crystallizable component of the sample and is proportional to the degree of crystallinity; (v) T_{onset} is the temperature at which crystallization begins.

MATERIALS AND METHODS

Four grades of polypropylene were used in the investigation:
(1) FINA 5042S, nucleated, MFI = 6.0 g/10 min;
(2) FINA 5060S, un-nucleated, MFI = 6.0;
(3) BASF 1184L, nucleated, MFI = 6.0;
(4) BASF 1100H, un-nucleated, MFI = 6.0.
For the trials with pigment as a nucleating agent, titanium dioxide based white pigment was mixed with both nucleated (BASF 1184L) and un-nucleated (FINA 5060S) in the following quantities; 0.5, 1, 2, 3 and 5%. The materials were thoroughly mixed by hand prior to compounding in a Killion KN-150 extruder to form pellets. Recycled material was produced by extruding FINA 5060S into pellet form. Four subsequent passes of this material through the extruder produced a further four generations of regrind. This regrind was compounded with virgin FINA 5060S at the following levels to produce pellets for DSC analysis (Perkin Elmer DSC7); 5, 10, 20 and 50%. For the DSC tests, a 10 mg sample of the required material was heated at 10°C/min from 20°C to 200°C, held for 2 minutes and then cooled at 10°C/min. On completion of the DSC test, the material was carefully removed from the DSC pan and etched with a permanganic acid solution prior to further examination using scanning electron mi-

croscopy (JEOL 6400). Melt flow index of virgin and recycled FINA 5060S was measured at 230°C/2.16 Kg on a Kayness 7053, while molecular weight measurements were carried out at Rapra Technology Ltd using gel permeation chromatography.

RESULTS AND DISCUSSION

THE EFFECT OF STANDARD NUCLEATING AGENTS

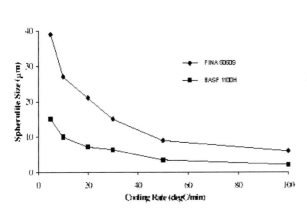

Figure 2. The influence of cooling rate on the spherulite size developed in melt crystallized samples of nucleated (BASF 1184L) and un-nucleated (FINA 5060S) polypropylene.

From the data in Table 1 it is obvious that the nucleated resins, FINA 5042S and BASF 1184L, crystallize from the melt at higher temperatures and at much greater rates, S_i, than the corresponding un-nucleated resins (FINA 5060S and BASF 1100H). The nucleated resins also have higher crystallinities (A/m) and narrower spherulite size distributions, Δw. In Figure 2 the difference in spherulite size between nucleated and un-nucleated materials is apparent. At a cooling rate (in the DSC) of 5°C/min the un-nucleated material has an average spherulite diameter of 39 microns while that of the nucleated material is 15 microns. Although this difference in spherulite size diminishes as the cooling rate increases to 100°C/min, the un-nucleated material spherulite is still twice the diameter of the nucleated grade at that condition. In annealing studies of both materials it was also found that the nucleated material achieved a maximum crystallinity after two minutes annealing at 155°C while the un-nucleated material took eight minutes to achieve its maximum value.

THE EFFECT OF PIGMENT

From Table 1 it is apparent that titanium dioxide based white pigment does act as a relatively efficient nucleating agent for FINA 5060S (un-nucleated) but has little effect in further enhancing the nucleation activity of BASF 1184L (nucleated). In both cases the greatest effect occurs at an addition level of 0.5% pigment. Increasing levels to 1.0% produces a further small effect, and thereafter the nucleation effect becomes static. Although the pigment particles do not nucleate the crystallizing melt to the same extent as the nucleating agent present in FINA 5042S, a fourfold increase in crystallization rate is nevertheless obtained, with a corre-

Table 1. Characterization of the activity of various nucleating systems

Material	T_p, °C	S_i	Δ_w, °C	A/m, J/g
FINA 5060S	109.90	1.10	4.60	93.00
BASF 1100H	115.80	2.80	5.40	97.80
FINA 5042S	126.40	5.70	2.50	104.30
BASF 1184L	120.80	4.30	4.00	103.60
FINA 5060S + 0.5% pigment	117.23	3.97	3.04	98.32
FINA 5060S + 1.0% pigment	118.95	4.25	2.90	99.40
FINA 5060S + 2.0% pigment	118.04	4.34	2.89	99.90
FINA 5060S + 3.0% pigment	118.46	4.40	2.79	100.01
FINA 5060S + 5.0% pigment	118.32	4.50	2.80	100.02
BASF 1184L + 0.5% pigment	122.97	4.68	4.26	104.99
BASF 1184L + 1.0% pigment	123.05	4.78	4.16	105.21
BASF 1184L + 2.0% pigment	123.36	4.89	4.12	105.32
BASF 1184L + 3.0% pigment	123.56	4.99	4.10	105.29
BASF 1184L + 5.0% pigment	123.65	5.01	4.11	105.28
FINA 5060S + 5% 1st generation regrind	111.20	1.75	4.43	93.90
FINA 5060S + 10% 1st generation regrind	111.89	2.10	4.35	94.30
FINA 5060S + 20% 1st generation regrind	112.56	2.50	4.30	94.80
FINA 5060S + 50% 1st generation regrind	112.69	2.55	4.25	95.30
FINA 5060S + 5% 3rd generation regrind	112.23	1.98	4.20	94.40
FINA 5060S + 10% 3rd generation regrind	112.99	2.40	4.10	94.80
FINA 5060S + 20% 3rd generation regrind	113.56	2.70	4.00	95.40
FINA 5060S + 50% 3rd generation regrind	113.68	2.73	3.89	95.60
FINA 5060S + 5% 5th generation regrind	113.21	2.40	3.80	95.01
FINA 5060S + 10% 5th generation regrind	114.65	2.90	3.60	95.94
FINA 5060S + 20% 5th generation regrind	115.50	3.30	3.50	96.50
FINA 5060S + 50% 5th generation regrind	115.70	3.40	3.45	96.60

sponding twofold reduction in crystallite size distribution. In Figure 3 the effect of pigment on spherulite size is shown. As in the case of nucleation efficiency there is little effect with 1184L. With FINA 5060S however, a 43% reduction in spherulite size is achieved in the presence of 0.5% pigment. This suggests that the pigment particles were acting as initial growth nuclei in the melt and successfully seeding the crystallization process. At higher levels of pigment the rate of decrease in spherulite size became less pronounced as the nucleating capability of the pigment particles became saturated.

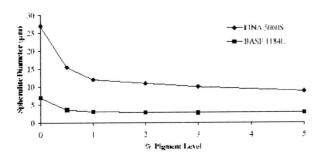

Figure 3. The influence of nucleating effect of varying levels of white pigment on the average spherulite size produced in nucleated (BASF 1184L) and un-nucleated (FINA 5060S) polypropylene.

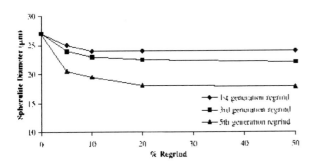

Figure 4. The effect of percentage regrind on spherulite diameter.

THE EFFECT OF RECYCLED MATERIAL (REGRIND)

The results in Table 1 indicate that incorporation of recycled material into virgin, un-nucleated FINA 5060S resin does indeed improve the crystallization process but to a much lesser extent than standard nucleating agents or titanium dioxide pigment. Fifth generation regrind at an addition level of 10-20% gives the best results in these trials. The weight average molecular weight of the resin reduced from 285,000 (virgin material) to 270,000 on the fifth pass through the extruder. This would indicate that a degree of chain scission took place which resulted in the presence of some low molecular weight chain break pieces in the regrind which in turn act as growth nuclei during the crystallization process. Figure 4 shows the effect of regrind on spherulite size with the best results again being obtained for fifth generation regrind at 10-20% addition levels.

CONCLUSIONS

1. FINA 5042S (nucleated) crystallizes at a temperature approximately 10°C higher, and at a rate five times greater than un-nucleated FINA 5060S. A similar (though lesser) effect is achieved in the BASF grades examined.

2. Spherulite size in un-nucleated material is very sensitive to melt cooling rates. It is therefore essential to have good control of cooling rates in such materials in order to achieve end product consistency.

3. The addition of 0.5% titanium dioxide based pigment to FINA 5060S (un-nucleated) is sufficient to increase its crystallization temperature by 7°C and its crystallization rate four-fold. Spherulite size also reduced from 27 microns to 15 microns at this level of addition.

4. Recycled material acts as a nucleation agent in polypropylene but is much less efficient than titanium dioxide pigment in this capacity unless higher generation regrind is utilized.

5. Finally, it should be apparent from the results of this investigation that un-nucleated polypropylene resin may be nucleated to varying degrees of efficiency by both pigment particles and recycled material. Since end product properties such as impact strength and elastic modulus are strongly dependent on both degree of crystallinity and on spherulite size it is imperative that the processor is aware of the nucleating potential of additives and that strict control of additive levels is maintained in order to guarantee product consistency.

ACKNOWLEDGMENTS

The authors wish to thank Fina Chemicals for donating the FINA 5060S resin used in this study. We also wish to thank Wilsanco Plastics Ltd for donating the remaining polypropylene resins. This work was supported by a grant from the Northern Ireland Teaching Company Scheme.

REFERENCES

1   A. K. Gupta, B. K. J. Ratman, J. *Appl. Polym. Sci.*, **42**, 297 (1991).

Relationship Between the Microstructure and the Properties of Rotationally Molded Plastics

M. C. Cramez, M. J. Oliveira
Dept. Eng. Polímeros, Universidade do Minho, 4800 Guimarães, Portugal
R. J. Crawford
School of Mechanical and Process Engineering, Queen's University, Belfast, N Ireland

ABSTRACT

The rotational molding process has characteristic features that make the microstructure of the molded plastic articles unique. The fact that the rotation speeds are slow (typically less than 10 rpm) results in low shear which promotes the development of textures that are free from orientation, but prevents the dispersion additives like pigments. Also, if the heating of the polymer is too severe, degradation may occur at the inner free surface, causing the spherulites that grow freely at that surface to be replaced by a non-spherulitic or a transcrystalline texture, depending on the extent of degradation. The polymer microstructure in the bulk, and also at the inner surface layer has a major influence on the mechanical properties of the molded material.

In this paper, work done on polyethylene and polypropylene rotational moldings will be presented to illustrate the influence of a number of factors: molding temperature, grinding and mixing conditions, type of pigment, anti-oxidant level, mold material and inner atmosphere. Polarized light, common light and DIC microscopy were used to study the microstructure of the moldings. The occurrence of degradation was assessed by FTIR spectroscopy and fluorescence microscopy. The rheological properties of the material after processing were determined with a cone and plate rheometer. The mechanical properties were measured by instrumented falling weight impact and tensile tests.

As a result of this extensive experimental program, it has been found that in rotational molding, polyethylene and polypropylene polymers show different degradation behavior. While in polypropylene the thermo-oxidative degradation causes mainly chain scission, in polyethylene crosslinking dominates. The use of increased amounts of antioxidant in the polymer, or the use of an inert atmosphere, delays the degradation but does not prevent it. It has also been observed that the thermo-mechanical effects caused by the mixing processes commonly used to add pigments increase the mechanical properties of polyethylene products. Finally, the nucleating activity of the pigment combined with the mixing process, has a major effect on the microstructure and on the mechanical properties of the final products.

INTRODUCTION

Rotational molding is used to manufacture one-piece hollow plastic articles.[1] It is currently the fastest growing sector of the plastics processing industry, with annual growth rates in the range 10-12%. Whilst rotational molding has many advantages such as low mold costs, seamless and stress-free moldings, controlled wall thickness distribution etc, it is characterized by no shear during shaping and slow cooling rates. These two factors lead to unique structural features in rotomolded products.[2-4] The absence of shear does not encourage good mixing of additives such as pigments, and the slow cooling promotes large spherulitic growth. On top of this, the longer cycle times and the presence of oxygen at the inner free surface of the moldings means that degradation processes can be initiated very quickly if the correct molding conditions are not used.[3,5]

Whilst there has been extensive work done on the mechanical properties of rotomolded products[5-8] and the heat transfer processes occurring during molding,[9,10] there has been relatively little work done on the unique microstructures that occur within rotomolded articles. This is surprising in view of the fact that the structures are very amenable to microstructural analysis. They display classic features in terms of spherulite geometries, and it is possible to relate mechanical properties to defects caused by the onset of thermal/oxidative degradation.

This paper examines the structures that occur in rotomolded polyethylene and polypropylene produced under a wide range of processing conditions. The effects of pigments are investigated as well as the mechanical and thermal processes that plastic powders typically undergo during preparation for rotomolding. In all cases, the mechanical properties of the molded articles have been related to the microstructure.

EXPERIMENTAL

In this investigation two types of polyethylene and one of polypropylene powder were used. A PE grade from DuPont, Sclair 8504, was used for the study of the effect of anti-oxidant concentration, atmosphere inside the mold, mold material and pigmentation. A PE grade from Borealis, NCPE 8644 was used for the study of the effect of the grinding temperature. The PP grade, BE 182B, was supplied from Borealis. The moldings, of approximately 3 mm in thickness, were produced using a shuttle type rotational molding machine and a three-arm machine. The oven, mold and internal atmosphere temperatures were monitored using a Rotolog system.[5] The effect of the heating rate was investigated by changing the mold material (aluminum, stainless steel and mild steel) and mold wall thickness.

The effect of grinding at three temperatures (40, 60, and 80°C) using a Wedco grinder, and mixing of pigments, having different nucleating capabilities, by turbo blending and extrusion was also investigated.

The moldings were characterized by several techniques: optical microscopy, differential scanning calorimetry, FTIR spectroscopy, tensile and impact testing and cone and plate oscillatory rheometry.

RESULTS AND DISCUSSION

EFFECT OF THE THERMAL TREATMENT

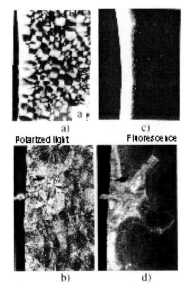

Figure 1. Cross-sections showing the inner free surface morphology. a,c - PE, b,d -PP.

The microstructure and the properties of polymers are very sensitive to the thermal treatment given to them during the molding stage. Measuring the maximum temperature of the atmosphere inside the mold during rotational molding permits the thermal history of the plastic to be monitored and related to structural variations. Polarized light microscopy of cross-sections of rotomolded articles show that when this temperature is too low, the samples have voids and the crystalline texture is spherulitic all through the thickness. If the heating is too severe and the temperature inside the mold goes above a certain value, the morphology changes at the free inner surface (Figure 1a and b). In the case of polyethylene, the changes of microstructure as the overheating increases are very clear. For moderate overheating, the size and perfection of the spherulites near the inside surface reduces and a inward growing layer of columnar type structure starts to appear. When the heating is too severe, the spherulitic texture disappears completely near the inside surface, and is replaced by a dark ribbon of material, apparently without any texture (Figure 1a) . For polypropylene, the overheating of the polymer did not prevent the crystallization of spherulites near the inner surface, but it increased their size and birefringence (Figures 1b). The increase in birefringence suggests that the lamellae cross-hatching of the type I spherulites, according to Padden and Keith classification[11] might have reduced.

Fluorescence microscopy, which enables the detection of double bonds in the polymer chains,[12] showed that the altered microstructure at the inner surface fluoresces (see Figures 1c

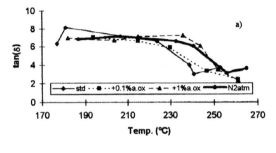

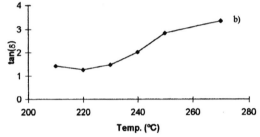

Figure 2. Effect of the processing temperature on the tanδ of inner surface material. a - PE, b - PP.

and d), indicating that the thermo-oxidative reactions took place there during the heating of the polymers. Further evidence of degradation, when overheating occurred, was given by FTIR spectra obtained from layers cut near the inner skin. Carbonyl, vinyl and hydroperoxide groups, that are common products from oxidative reactions of polyolefins,[13,14] were identified in the spectra.[3] The use of a nitrogen atmosphere and an increase in the concentration of antioxidant did not prevent the above morphologies from being formed but raised the temperature at which they occurred.

The rheological results plotted in Figures 2a and b for polyethylene and for polypropylene, respectively show

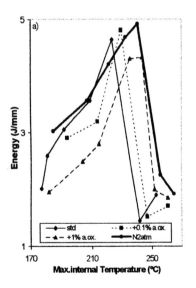

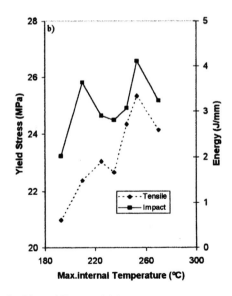

Figure 3. The effect of maximum inner temperature on the impact strength of the moldings. a - PE, b - PP.

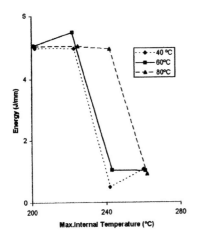

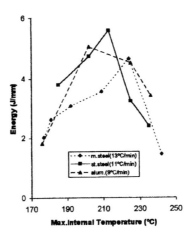

Figure 4. Effect of heating rate on the optimum processing temperature of PE.

Figure 5. Effect of the grinding temperature on the optimum processing temperature of PE.

that the degradation behavior of the two polymers differs. In the case of polypropylene, the tan(δ) increases from fairly low temperatures, while for the polyethylene the tan(δ) is almost constant up to a certain processing temperature and from then on a steadily decrease with temperature is observed. These results indicate that during the degradation of polypropylene, chain scission predominates over crosslinking for all the processing temperatures while in the case of polyethylene, a clear unbalance towards cross-linking occurs above a certain temperature.

The mechanical properties of the samples are very sensitive to the maximum inner air temperature (see Figures 3a and b). The best properties were obtained with the samples that were heated up to a temperature that facilitated the diffusion of air out of the melt without causing degradation at the inner surface. The optimum inner air temperature varies with a number of factors. The use of nitrogen or a higher concentration of antioxidant delayed the degradation process and shifted the optimum temperature to higher values (Figure 3a). A similar effect is observed when the heating rate increases (Figure 4). This effect is associated with the time of exposure of the polymer to high temperatures that decrease when the heating rate is increased. The temperature of grinding of the polymer before molding also affects the optimum process temperature (see Figure 5). A grinding temperature of 80°C helps the polymer to withstand higher processing temperatures (242°C) without losing impact strength. The reason for this behavior may be the annealing effect of the grinding at a higher temperature that increased the crystallinity and consequently the amount of energy absorbed by the polymer during the melting.

EFFECT OF PIGMENTATION AND MIXING METHOD

The turbo blending of the polymer further changes its crystallization behavior. When Figures 6a and b are compared, one see that, in the absence of pigment, rows of small spherulites have grown at the polymer particles boundaries. This might be due to the thermo-mechanical action of the rotating blades that somehow modifies the structure of the molecules around the particles. The extrusion mixing reduced the size of the spherulites (Figure 6c), probably due an orientation memory of the extruded material that induced nucleation during the molding stage.

The polarized light and bright field micrographs shown in Figure 7 illustrate the effect of the nucleating ability and mixing efficiency of the pigments into the microstructure of polyethylene samples. The nucleating effect is more noticeable in the turbo-blended samples, because as the pigment is spread around the original polymer particles, transcrystalline textures were formed at their boundaries. The nucleating properties of the pigment also have an effect on the microstructure of the extruded samples (Figure 7d), although the mechanism of this is not yet fully understood.

It is apparent from the DSC thermograms shown in Figure 8 that the extrusion process prior to molding increases the enthalpy of fusion and the crystallization temperature of the

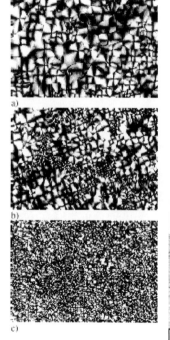

Figure 6. Effect of mixing on the microstructure of virgin PE (polarized light). a - original material, b - Turbo blended, c - extruded.

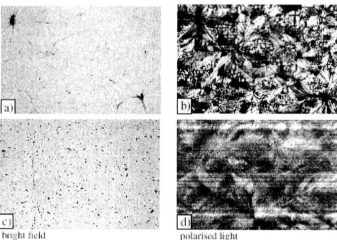

Figure 7. Effect of the nucleating activity of pigments on the microstructure of PE. a, b - Turbo blended material, c, d - extruded material.

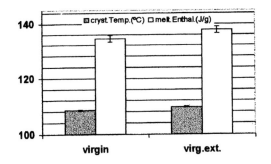

Figure 8. Effect of extrusion on the thermal properties of PE.

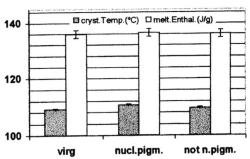

Figure 9. Effect of pigmentation on the thermal properties of Turbo blended PE.

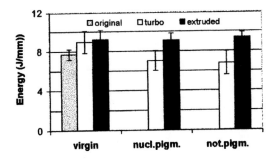

Figure 10. Effect of pigmentation and mixing on the impact strength of PE.

polymer. As expected from other works with nucleating additives[15,16] a similar increase was observed when a nucleating pigment was added by turbo blending (Figure 9).

It is an important observation that the microstructural modifications caused by the extrusion and turbo-blending in the unpigmented samples result in higher impact strengths than those observed in the original polymer (see Figure 10). This improvement in properties disappears if a pigment is mixed by turbo blending, probably due to the poor distribution capability of this technique.

CONCLUSIONS

In this work, the microstructure of rotationally molded samples has been related to the processing conditions and mixing of additives. This helped to explain the variations in mechanical properties. The following main conclusions can be drawn:

1. The maximum temperature reached by the air inside the mold is the more adequate parameter to control the process. It has a clear relationship with the microstructure and the properties of the moldings. Optimum values of these properties correspond to moldings free from voids and from degradation at the inner surface.

2. In rotational molding, polyethylene and polypropylene show a different degradation behavior. While in polypropylene the thermo-oxidative degradation causes mainly chain scission, in polyethylene crosslinking dominates.

3. The use of increased amounts of antioxidant, an inert atmosphere or of higher heating rates delays the degradation. Grinding the polymer at higher temperatures (80°C) seems to have a similar effect.

4. The thermo-mechanical action of the mixing processes, commonly used to add pigments, affects the microstructure and improves the mechanical properties of polyethylene moldings.

5. The nucleating activity of the pigment, combined to the mixing process, has a major influence on the microstructure and properties of the final products.

ACKNOWLEDGMENTS

This work was supported by JNICT, INVOTAN and The British Council. The authors are thankful to Dr. J. A. Covas for valuable discussions about the rheological results.

REFERENCES

1   R. J. Crawford, **Rotational Molding of Plastics**, *John Wiley and Sons Inc.,* (1992).
2   M. C. Cramez, M. J. Oliveira, R. J. Crawford, Thirteenth PPS Annual Meeting, New Jersey, USA, June 1997.
3   M. J. Oliveira, M. C. Cramez, R. J. Crawford, *J. Mat. Sci.*, **31** (1996) 2227.
4   M. J. Oliveira, M. C. Cramez, R. J. Crawford, P. J. Nugent, Europhysics Conf. Macrom. Phys., Prague, Czech Republic, July 1995.
5   R. J. Crawford, P. J. Nugent, *Plastics, Rubber and Composites, Processing and Application*, **17**, 1 (1992).
6   M. A. Rao, J. L. Throne, *Pol. Eng. Sci.*, **12**, 4 (1972) 237.
7   J. L. Throne, M. S. Sohn, *Adv. Polym. Techn.*, **9**, 3 (1989) 181.
8   K. Iwakura, Y. Otha, C. H. Chen, J. L. White, *Int. Polym. Processing*, **4**, 3 (1989) 76.
9   J. L. Throne, *Polym. Eng. Sci.*, **16**, 4 (1976) 192.
10   J. L. Throne, *Polym. Eng. Sci.*, **12**, 5 (1972) 335.
11   F. J. Padden, Jr, H. D. Keith, *J. Appl. Phys.*, **30**, 10 (1959) 1479.
12   D. A. Hemsley, R. P. Higgs, A. Miadonye, *Polym. Commun.*, **24** (1983) 103.
13   M. Iring, Tüdós, *Prog. Polym. Sci.* **15** (1990).
14   H. Hinsken, S. Moss, J.-R. Pauquet, H. Zweifel, *Polym. Degrad. Stability*, **34** (1991) 279.
15   F. L. Binsbergen, *Polymer*, **11** (1970) 252.
16   M. Burke, R. J. Young, J. L. Stanford, *Plast. Rubb. Comp. Proc. Appl.*, **20** (1993) 121.

Colored Engineering Resins for High Strain/Thin Walled Applications

Bruce M. Mulholland
TICONA

ABSTRACT

As part designers push the limits of strain requirements on levers and snap-fits, or continue to reduce size and cost by designing thinner and thinner sections, processors often are faced with breakage problems with colored resins. The problem can be unexpected if the new part design was prototyped in natural, or the processor had been using the colored resin in an existing design without problems. This paper discusses two main causes of breakage problems with engineering resins — colorant selection and colorant dispersion.

INTRODUCTION

Much has been documented concerning the affect of colorants, primarily carbon black and titanium dioxide, on the properties of polymers. Polymers generally studied include polyolefins and ABS. Properties talked about are usually viscosity of color concentrates, color development and consistency, or occasionally mechanical properties. Unfortunately, little has been documented on the affect of colorants on engineering resins specifically, and their influence on mechanical properties in particular.

Indeed, it is the properties of the engineering resin which most likely led to its selection in the first place, so the influence of colorants on them is of utmost importance. Molders and part designers generally understand that mechanical property loss can be expected from the addition of colorants. However, problems arise when they grossly underestimate their effect. A recent informal survey of part designers revealed that they typically plan for a 5% reduction in mechanical properties due to colorants. Depending on the resin and colorant selection, this number can easily exceed 30%! What's more, the drive towards thinner and thinner wall sec-

tions to reduce part weight further aggravates this problem. All of this leads to questions (and frustration) from molders and designers such as:
- "Why do the black parts occasionally break?"
- "Why are the blue parts undersized?"
- "Why does not the red color work in this tool when it worked in the old design?"

and, of course, the most important question:
- "What are you the resin or colorant supplier going to do to fix the problem?"

In this paper, several common problems observed with colored engineering resins are discussed. These problems include:
- Unfilled resin part breakage
- Glass-reinforced resin part breakage
- Colorant induced shrinkage differences

This discussion includes explaining the cause of the problems and some suggested solutions. Each problem area is discussed in the sections that follow.

UNFILLED RESIN PART BREAKAGE

Unfilled resins for this discussion are any engineering resin not reinforced by mineral or fiberglass. These would include neat resins, impact modified resins, UV stabilized resins, and unfilled alloys to name a few. The problem typically observed when coloring these resins is that the part is no longer tough enough when compared to parts molded from natural resin. Part breakage or cracking may occur in cantilever type locking beams, snap-fit tabs or any other design accessory which requires the molded part to deflect under load.

The root cause for a breakage problem attributed to the colored unfilled resin is usually related to colorant dispersion issues. This assumes that the parts are molded to the proper size to rule out increased shrinkage discussed later in this paper. Undispersed or less than optimally dispersed colorant can act as stress concentrators reducing toughness. In particular, if these relatively large agglomerates of undispersed pigment end up in areas of the molded part where walls are thin or stresses are high, part breakage may occur.

Colorants are purchased from suppliers in dry powder form. These powders are in the form of agglomerates which are many primary, individual particles of the pigment fused together. Inorganic pigments such as titanium dioxide or iron oxide have primary particle sizes of 0.2 to 5 microns. Organic pigments have smaller particle sizes of 0.01 to 0.1 micron and include phthalocyanine and quinacridone pigments. Both types of pigments can have agglomerates of several hundred microns in the dry powder form. It is the job of the color compounder to reduce these agglomerates to primary or near primary sizes.

Colorant dispersion is typically achieved using a two step process. The first step involves dry mixing the powdered colorant and resin in a high intensity mixer to reduce ag-

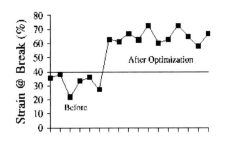

Figure 1. Strain at break of black polyester before and after compounding optimization.

glomerates. The dry mix is then compounded using any number of techniques including single or twin screw extrusion, batch or continuous mixers, or even roll mills. The end result is pigment dispersed and encapsulated in the polymer.

The degree of dispersion is key to preserving the properties of the unfilled engineering resin. Dispersion quality can be determined by microscopy which is time consuming and costly. A practical test to measure dispersion quality is elongation or strain at break. This test correlates well to field failures and is less costly to perform than microscopy. Using a practical measure like strain at break allows the resin supplier to optimize their process to improve dispersion. An example of this is shown in Figure 1. This data shows improvement in strain at break of a polyester resin achieved by optimizing the compounding process to disperse carbon black.

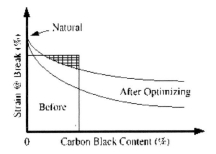

Figure 2. Strain at break vs. carbon black content.

With agglomerates present, strain at break will generally decrease with increasing colorant concentration as shown in Figure 2. If the strain requirement in the part is below the colored material's maximum strain as shown by the solid line in Figure 2, there will not be a breakage problem in the part (at least not attributed to color or material issues). However, if the strain requirements are changed due to reducing wall thickness or adding an accessory like a snap-fit tab, the localized strain requirements may exceed the colored material's capability. This situation, shown by the hatched area in Figure 2, causes the breakage problems to occur.

There are two basic ways to minimize the effect of pigment concentration on strain at break thereby making the colored material behave more like natural resin. The first is simply color selection. In general, organic pigments are more difficult to disperse than inorganic pigments. This is due to their smaller primary particle size. The smaller size creates much higher surface area requiring significantly more energy to break up the agglomerates of pigment. A typical dispersion of organic pigments in engineering resins may contain some agglomerates in the order of 20 up to 100 microns depending on the dispersion quality. In contrast, inorganic pigments are much easier to disperse with less propensity for agglomerates to be present in the colored resin. Therefore, colors that can be developed using inorganic colorants will generally have less agglomerates present and thus higher strain at break values. Colors in this family include whites, off-whites, light grays and pastel colors. Colors to avoid then include dark blues, maroons, dark browns and even black as these typically require higher levels of the more intense organic pigments to produce these shades.

While color selection is generally helpful in minimizing pigment agglomerates, it is not very practical, especially if the customer requires a black or dark blue part. Thus, optimum colorant dispersion must be achieved to minimize the effect of pigment concentration on strain at break. This is the second and most important solution. As previously mentioned, dispersion is typically achieved using a two step process. The compounding step is generally thought to be the most important step and is often the target for process optimization work. In fact, the data presented in Figure 1 is the result of compounding optimization work to improve dispersion of carbon black in polyester. Even with that significant step change shown in Figure 1, the improved product still exhibited a greater than 3,000 ppm breakage rate in an automotive connector application.

Optimum pigment dispersion must include process development of the dry mixing step as well. High intensity mixers are typically used for this dry mixing step. Controllable variables include mixing blade configuration, deflector blade position (if used), RPM and cycle time. Work was recently performed to optimize this dry mixing process for the carbon black product mentioned above. Combining the results of that work with the optimized compounding process yielded a polyester black product with strain at break values equivalent to natural resin. Using this new technology in the automotive connector application dropped breakage associated to the material to zero!

In summary, breakage problems in parts molded from colored, unfilled engineering resins may be attributed to pigment agglomerates present in the resin. The problem is more prevalent in darker colors where the use of organic pigments is required. The way to solve this problem is to maximize dispersion of the pigments. Optimization should include all points in the process such as dry mixing and compounding. The result can be strain at break values of colored materials equivalent to natural resin. Still, parts can break for other reasons generally related to molding conditions or mold design which need to be investigated as well. Finally, if the strain requirements exceed that of the natural resin as illustrated in Figure 3,

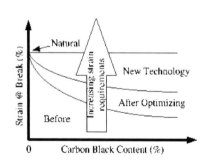

Figure 3. Property optimization.

a different base material would need to be selected or significant design changes made to reduce strain requirements.

GLASS-REINFORCED RESIN PART BREAKAGE

This section discusses the breakage problems attributed to colorants in glass-reinforced engineering resins. These resins contain fiberglass and other additives to increase tensile, flexural and Izod impact properties. Molded part breakage may occur as cracks or complete breaks generally in thin walled areas. The reason for failure in these resins is totally different from the unfilled resins discussed above. In glass-reinforced resins, pigment dispersion and agglomerates are usually not a problem. The main causes of breakage are either the colorant interfering with the resin/glass bonding, or abrasion of the fiberglass by the pigment to reduce fiber length and thus strength. The solution is for the colored resin supplier to understand this phenomenon and only formulate colors in glass-reinforced resins using colorants which have little to no effect on mechanical properties. As an example, three colors, a red, yellow and white, can be formulated using two different colorant systems for each. The resin is 30% glass-reinforced polyester. With each color, one can formulate poorly and experience up to a 30% reduction in strength. Or, these same three colors can be formulated using appropriate technology to preserve strength at virtually 100% of the natural resin.

COLORANT INDUCED SHRINKAGE DIFFERENCES

Another source of problems and frustration with colored resins is associated with shrinkage differences between various colors, and colors versus natural. These shrinkage differences can cause problems such as undersized/oversized parts, warped parts, cracked/broken parts, and processing problems like cavity or core sticking. The root cause is that some colorants, mainly organic pigments, provide crystallization nuclei which initiate crystal growth in the polymer. Crystallization rate changes as well as morphological changes will result in different shrinkage rates for these colors. Inorganic pigments typically show little effect on shrinkage; many organic pigments induce up to 20% more shrinkage than the natural resin. Work to characterize shrinkage versus various colorants was completed in conjunction with Penn State Erie Campus.[1] Data is shown in Table 1.

When shrinkage issues are present, there are two basic solutions: (1) reformulate the color to reduce the shrinkage; or (2) employ a different set of molding conditions for the higher shrink colors. Molders understandably prefer option 1 so that all colors in a particular resin mold using the same conditions. While this is a worthy objective, we must not loose sight that twenty different colors in Resin A is no different than the same color in twenty different Resins A through T — they both represent twenty different formulations. And twenty different formulations would typically require twenty different sets of molding conditions!

Fortunately, things are not as bad as this. Looking at the shrinkage values for colorants in Table 1, it appears that probably no more than two sets of conditions would cover all of these

Table 1. Linear shrinkage data of colorants in PBT

Colorant type	Color index	Concentration, %	Shrinkage, mm/mm	Shrinkage vs. natural, %
None (natural)	n/a	0.00	0.0176	-
Inorganic				
Titanium dioxide	White 6	1.00	0.0181	2.8
Mixed metal oxide	Brown 24	1.00	0.0187	6.3
Red iron oxide	Red 101	1.00	0.0189	7.4
Organic				
Azo yellow	Yellow 183	0.10	0.0193	9.7
Azo yellow	Yellow 183	0.50	0.0195	10.8
Perylene red	Red 178	0.10	0.0204	15.9
Perylene red	Red 178	0.50	0.0207	17.6
Quinacridone red	Red 202	0.10	0.0207	17.6
Quinacridone red	Red 202	0.50	0.0210	19.3
Phthalocyanine green	Green 7	0.01	0.0201	14.2
Phthalocyanine green	Green 7	0.010	0.0206	17.0
Phthalocyanine green	Green 7	0.50	0.0208	18.2
Phthalocyanine blue	Blue 15:3	0.01	0.0183	4.0
Phthalocyanine blue	Blue 15:3	0.10	0.0203	15.3
Phthalocyanine blue	Blue 15:3	0.50	0.0205	16.5
Phthalocyanine blue	Blue 15:4	0.01	0.0186	5.7
Phthalocyanine blue	Blue 15:4	0.10	0.0205	16.5
Phthalocyanine blue	Blue 15:4	0.50	0.0205	16.5
Carbon black	Black 7	0.30	0.0192	9.1

colors. One set of conditions would be required for natural and inorganic type colors (whites, grays, tans and pastels), and another set for chromatic colors requiring organic pigments such as bright reds and dark blues. What's more, depending on the part design and dimensional tolerances, for a majority of applications all twenty colors can be molded using the same set of conditions. It is usually when mold design and/or the molding condition window are marginal do different colors require their own set of processing conditions to meet dimensional tolerances.

Formulating to minimize shrinkage differences has its drawbacks as well. The first is metamerism. Metamerism occurs when two color samples (a standard and batch) match under one light source but no longer match when the light source is changed. This is important in automotive interiors, for example, where the OEM wants all of the materials in the vehicle to match in color under a variety of lighting conditions. Metamerism is avoided if both samples contain the same colorants at similar concentrations. If, for example, an automotive gray was mastered by the OEM containing a small amount of phthalocyanine green, the resulting gray in Resin A will shrink significantly more than other grays not containing that green pigment. The color supplier could formulate the gray without the green pigment to reduce shrinkage, but metamerism would likely be present and may be objectionable.

The other drawback with formulating around shrinkage issues is that the desired color may no longer be achieved. The best example is a dark royal blue. Virtually the only pigment to achieve such a color is phthalocyanine blue. The data in Table 1 shows that higher shrinkage is likely to occur. Using an inorganic blue like ultramarine or cobalt may not yield the necessary depth of color making achieving the dark royal blue impossible. In these cases, reformulating the blue is not an option and molding conditions must be used to control dimensions, or the color must be changed to one that can be achieved using a low shrink formulation.

CONCLUSIONS

Hopefully this paper met the objective of presenting several common problems observed with colored engineering resins, explaining the cause and offering some potential solutions. Of course, this cursory look just brushes at the surface but should provide some guidance as to what to look for when problems occur. The writer must also clearly point out that all of these problems may be related to other factors not attributed to the color at all. Breakage problems can be mold design related, processing condition related, or simply material selection. While you should not blame all the problems on the color, do not rule it out too quickly either! This paper should help to decide whether the problem could be attributed to the color or not.

REFERENCES

1   Neubert, C., Effect of Colorants on the Shrinkage of Polyester, Society of Plastics Engineers, Tech. Papers ANTEC '94, Vol. XXXX, p. 3610 - 3613.

Feasibility of Automotive Coatings Designed for Direct Adhesion to TPO Materials

Michelle Mikulec
Ford Motor Co, PO Box 623, Dearborn, MI 48121, USA

ABSTRACT

The discovery, that adhesion to untreated non-polar TPO could be achieved directly through olefinic color coats based on low viscosity functional liquid polymers, might be considered the breakthrough for the TPO industry. Feasibility of the olefinic color coats at bumper plants is also discussed.

INTRODUCTION

The industry thinking for the past 30+ years required the non-polar, olefinic TPO to conform to the polar coating systems. The new paradigm suggests that the coatings should be designed for the olefinic, non-polar TPO, therefore eliminating surface pretreatment. The novelty of this approach is Shell's discovery that the low viscosity functional "Kraton Liquid™ Polymer" (hydrogenated polybutadiene diol, MW 3000) could be formulated into production of color coats, thus eliminating the need for the TPO pretreatment process (Figures 1 and 2). Examples of the surface treatments to enhance the adhesion between the TPO substrate and the color coat are:

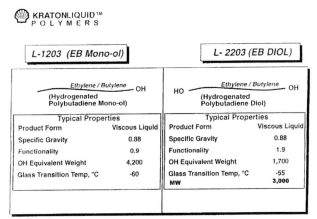

Figure 1. Properties of materials

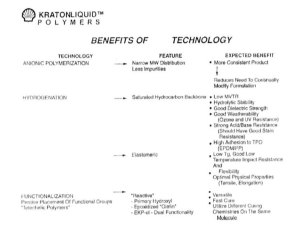

Figure 2. Benefits of technology.

adhesion promoter, followed by flame and plasma treatments; another approach is to use directly paintable TPO.

EXPERIMENTATION

The assessment of Shell's olefinic color coats was done in February 1996. A preliminary routine evaluation of Shell's unpigmented olefinic color coats combined with the production clear coats and applied over the production TPO substrate showed extraordinary adhesion properties. Long term testing and evaluation of mar, gouge and scratch has not yet been completed.

MANUFACTURING FEASIBILITY OF PIGMENTED COLOR COATS BASED ON LOW VISCOSITY OLEFINIC POLYMERS

The difference in the TPO pretreatment processes and their quality control methods lies in two areas: Can the use of a certain pretreatment eliminate a manufacturing step? Secondly, what type of quality control methods are used to quantify the polarity/modification of the TPO surface and their simplicity of use and therefore cost? The following is a brief comparison of the pre-treatments, their quality control methods and their actual or hypothetical adhesion mechanisms (Figure 3).

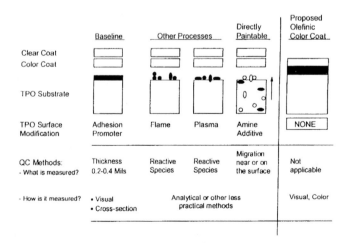

Figure 3. Are the olefinic color coats feasible in manufacturing?

ADHESION PROMOTER - BASELINE

The TPO parts are power washed and chlorinated polypropylene adhesion promoter, AP, is applied to the surface. To achieve sufficient paint adhesion, it is required that the AP thickness range is between 0.2-0.4 mils. Then the color coat, either 1K or 2K is applied, flashed and followed by the application of a 1K or 2K clear coat. The entire system is cured in the paint oven usually at 121°C for 30 minutes. QC method: TPO surface: 1/ visual; Painted part: 1/ AP thickness ranging from 0.2-0.4 mils is measured *via* cross-section under the microscope, 2/ cross-hatch adhesion, 3/ other tests.

FLAME TREATMENT

The power washed TPO parts pass under a robotic oxygenated butane or propane "flame arm". The oxidizing flames (flame plasma) change the chemical make up of the TPO surface. The basis for this change is a working knowledge of the combustion reaction of the hydrocarbon gas defined as:

$$Cn\ Hm + O_2 = \textit{free radical of which 44 are known} = CO_2 + H_2O + Heat$$

In the treatment of plastics, the free radicals generated in the reaction chemically modify (ionize) the surface of the TPO thus allowing the color coat to attach itself to the polar TPO surface. The rest of the process is the same as in the "Baseline" above. QC method: TPO surface: 1/ FTIR 2/ surface tension; Painted part: 1/ cross-hatch adhesion 2/other tests.

PLASMA TREATMENT

A TPO part (no power wash required) enters a vacuum chamber and gases (oxygen or nitrogen) are introduced into the chamber. The gases are ionized, and begin to glow: plasma is thus formed. The plasma treatment of the TPO surface is a result of several interactions of the plasma with the polymers in the TPO; it removes adsorbed atoms and weak layers from the polymer surface, creates functional groups (radicals) which increase surface wettability and paint adhesion. The vacuum is released and the color coat is applied directly on the now polar TPO surface to initiate the adhesion. The rest of the process is the same as in "Baseline" above. QC method: TPO surface: 1/ FTIR, 2/surface wetting; Painted part: 1/cross-hatch adhesion, 2/ other tests.

DIRECTLY PAINTABLE TPO

An amine additive is added to the TPO formulation at the compounder. The integrity of this amine additive system depends upon being well dispersed into the TPO compound by the compounder and maintained during the injection molding of the TPO part. The part is then power washed, color and clear coats are applied on the now modified surface which will achieve adhesion by the presence of functional groups near or on the TPO surface. It is assumed that the adhesion between the color coat and the TPO depends on uniform migration of the amine additive to or near the surface. There may be other adhesion mechanisms occurring simultaneously. QC methods: TPO surface: FTIR; Painted part: 1/ cross-hatch adhesion, 2/other tests.

THE OLEFINIC COLOR COATS BASED ON LOW VISCOSITY FUNCTIONAL POLYMERS

The TPO part is power washed and the olefinic color coat is applied on the untreated, non-polar TPO surface. The ability of the "Kraton Liquid Polymer" to be formulated into a production color coat is a recent development and the adhesion mechanism is believed to be based on diffusion of very similar materials into each other with the aid of heat and solvents during the paint bake cycle. There may be other adhesion mechanisms occurring simultaneously. QC method: TPO surface: 1/visual, on-line identifies the color of the pigmented color coat ranging in thickness between 0.9-1.2 mils; Painted part: 1/ cross-hatch adhesion, 2/ other tests.

BENEFITS

The following benefits may be realized:

1) Elimination of a manufacturing process step: It may be more cost effective to use olefinic color coats with their direct adhesion to the non-polar, untreated TPO surface than to use an extra manufacturing step to pre-treat the TPO surface with AP, flame or plasma.

2) Low cost, simple on-line quality control method:

a) The color of the olefinic pigmented color coat when applied directly on the untreated TPO surface will be visible to the bumper paint line operator. The color of the bumper surface may be used as an indicator of the presence of the color coat, thus ensuring uniform adhesion on the entire surface. Control of the olefinic color coat thickness may not be as critical as it is for the AP thickness, range 0.9-1.2 mils or 0.2- 0.4 mils respectively, and thus allows for a wider process window with the olefinic color coats.

b) It is assumed that it is simpler and therefore less costly to identify visually the color of the color coat on TPO surface, thus adhesion, than to depend for adhesion through uniform migration of the amine additive to or near the surface as expected by design in the directly paintable TPO.

3) Minimization of Compounded effects of "errors" during manufacturing: It is estimated that the compounded effects of "errors" when olefin color coats are used may be lower than in above mentioned pre-treatments and their quality control methods. This notion is based on the capability of the olefinic color coats to achieve direct adhesion to the untreated

TPO with a minimum of manufacturing process steps as compared to other, more involved pretreatment processes.

CONCLUSIONS

Based on this comparison of TPO pretreatment processes, their quality control methods, including actual (a, b, c) or hypothetical (d, e) knowledge of their adhesion mechanisms and their benefits, it was concluded that the low viscosity olefinic polymers may be feasible to be incorporated in the production pigmented color coats. A rigorous evaluation is essential before their application can be adapted in the automotive manufacturing plants, and the benefits realized.

REFERENCES

1   D. J. St. Clair, Coating Resins Based on Melamine Cured Polyolefin Diol, TPO in Automotive '96.
2   K. McNeal, Low viscosity functional liquid polymers: A new approach for adhesion to thermoplastic polyolefin (TPO) substrates, TPO in Automotive '96.
3   Aerogen Co., Gas Flame Plasma Treating for Automotive Products, Product Literature, 1995.
4   Balzer Co., The industrialization of the cold plasma process for the pretreatment of polypropylene Bumpers, Company Report, 1995.

Infrared Welding of Thermoplastics. Colored Pigments and Carbon Black Levels on Transmission of Infrared Radiation

Robert A Grimm, Hong Yeh
Edison Welding Institute

ABSTRACT

This project evaluated the heating characteristics of infrared energy from a quartz-halogen lamp (maximum output at 0.89 m) on some colored polymers.

One polymer was acrylonitrile-butadiene-styrene, ABS, pigmented with different colors: blue, green, orange, yellow, and red. Studies of the effect of carbon black on infrared absorption were made by making mixtures of polyethylene with different levels of carbon black.

The polymers were made into films. Power and temperature measurements were taken by locating the film between an infrared heater and measuring the amount of energy that was transmitted through the film.

Polymers pigmented with carbon black showed the least transmission (most absorption) and only very low levels of carbon black were needed to make them opaque.

Examination of colored films of ABS showed increased transmission as the colors progressed through the spectrum from red to blue. That is, red absorbed more energy relative to the other colors even though the spectral output was weighted more heavily in the red part of the spectrum.

INTRODUCTION

Infrared welding has been characterized as unpredictable since different polymers or formulations have been observed to heat at widely different rates under similar conditions. A previous reference reported a significant difference in absorption between thermoplastics that contained no carbon black and a similar material with carbon black levels around 0.5 percent,[1] but no intermediate levels were examined to determine the minimum levels at which absorption became essentially total.

Polymers that have pigments with other colors of the spectrum and/or can scatter light might also be expected to show differences in absorption of infrared energy. This is because the flux density of the radiation from a normal quartz-halogen source is greater on the red side of the spectrum than it is at the blue end of the spectrum.

Quartz-halogen lamps, with filament temperatures in the range of 3000°C, heat predominantly through radiation. The output from filament or thermally-heated sources can span a range of wavelengths but the maximum output is at a wavelength predicted by Wein's Law:[2]

$$\lambda_{max} T = 0.2898 \times 10^4 \ \mu m \ K$$

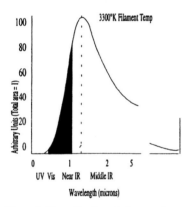

Figure 1. Normalized spectral output from a quartz-halogen lamp (shaded area is the UV-visible range (300 to 720 nm).

While most of the radiation from a quartz-halogen source is emitted at 0.89 μm, small amounts are emitted at wavelengths as short as 0.3 μm (ultraviolet rays). There is a substantial visible component to this light and there is relatively more red than blue in it. A distribution curve is shown in Figure 1, but it should only be considered approximate at this time.

Pigments, fillers, coatings, and other components of polymer formulations affect what happens to incident infrared radiation. They will strongly affect the ability of polymers to absorb the IR radiation, and can lead to various amounts of reflection and absorption (and transmission). This study aimed to study the absorption characteristics of IR energy by different polymers, by variously-colored ABS materials, and by polymers with different levels of carbon black. This information should provide practical guidelines and process understanding when infrared welding is being used.

EXPERIMENTAL

DESCRIPTION OF EQUIPMENT AND PROCESS

Films of colored polymers were obtained by disassembling floppy diskettes, ABS, of different colors and using segments of the 0.25-mm-thick walls. Red, orange, yellow, green, and blue pieces were examined with thicknesses of 0.25 and 0.5 mm (two layers).

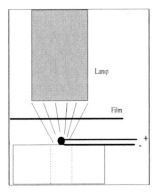

Figure 2. Experimental setup.

The polyethylene films with various levels of carbon black were prepared by mixing various ratios of black polyethylene (0.2 percent C) with natural polyethylene (w/w) and pressing them between Kapton films in a platen press. The film was cut, re-stacked, and re-pressed about ten times.

After these multiple pressings, uniform films were obtained in thicknesses of approximately 0.25 and 0.5 mm. Films were prepared with carbon black levels of 0.033, 0.05, 0.067, 0.1, 0.133, and 0.15 percent carbon.

Absorption behavior was inferred by measuring the amount of energy that was *transmitted* through the polymer films. The configuration of the lamp, film, and measuring devices is shown in Figure 2. Since the output from the lamp was constant, the greater the transmission, the less the absorption. This type of arrangement provides qualitative information that can be used for comparative purposes. Prolonged exposure led to melting and decomposition of the films, so care was used to expose the samples for lengths of time where they were not damaged.

The amount of transmitted energy was measured with a fixed thermocouple. Output was recorded with a computer-controlled data acquisition system operating with a sampling rate of 10 Hz. The thermocouple was positioned on a block of white acetal polymer (polyoxymethylene) to avoid any charring in the region of the sensor since this can have a significant effect on temperature readings.

The spot heating lamp was a custom-built, MR16, quartz-halogen type (General Electric EXS with a focal length of 4 cm and a smooth, aluminized reflector) operating at filament temperature around 3000°C. At the focal plane, the spot heater delivered a flux density in excess of 140 W/cm^2. At these temperatures, the maximum output occurred at a wavelength around 0.89 μm.

Infrared radiation is not visible to the human eye, but these lamps emit visible light of considerable intensity along with small amounts of ultraviolet light. For this reason, protective, dark green glasses were worn during all tests and EWI recommends this as a standard safety practice.

RESULTS AND DISCUSSION

EFFECTS OF CARBON BLACK LEVEL

Carbon black pigments found in black polymers represent an almost perfect absorbing material. The work reported here aimed at quantifying how light transmission depended on the level of carbon black that is present in a polymer.

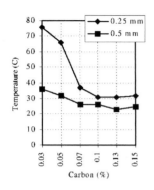

Figure 3. Effect of carbon black level on transmission through polyethylene.

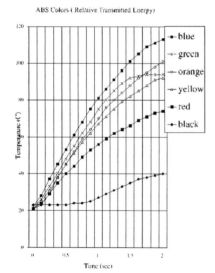

Figure 4. Transmission vs. color of ABS sheets.

Figure 3 shows the findings. The thin film of 0.25 mm showed a definite S-shaped variation, going from transparency to nearly complete absorption as the carbon black level went from 0.03 to 0.07 percent. If polymers are formulated with carbon black levels in this range, their welding behavior can be expected to be highly sensitive to tiny variations in formulation.

The thicker film, at 0.5 mm, shows much less of a transition in the range of 0.03 to 0.07 percent carbon. At this thickness, the effective amount of carbon black is increased, essentially doubled, and transparency will require even lower levels. These conclusions have been confirmed in studies conducted since this work was completed.

EFFECTS OF COLORED PIGMENTS ON ABSORPTION

ABS samples were cut from the walls of colored floppy disks and used as single and double layers for evaluation of the effects of color on absorption. It was assumed that the polymer formulations were the same, except for the color, thus isolating the effects of part color. Heating due to the near infrared wavelengths should have remained constant.

In summary, absorption was strongest for the red samples, weakest for the blue, and decreased as the colors changed from yellow to orange to green (Figure 4). Examination of the output spectrum from a quartz-halogen lamp shows that it has a higher flux density at the red end of the spectrum rather than the blue since output goes to zero just beyond the blue or violet end of the spectrum.

Materials have a certain color because they absorb all colors from the spectrum and re-emit the observed color. Thus, the red sample should be re-emitting the red color that constitutes the largest part of the visible output. However, the temperatures sensed with the red samples are lower than for other colors. Thus less energy is transmitted through the red ABS than through any other color.

The blue polymer is re-emitting the blue color which constitutes only a small part of the visible output. However, it results in a higher temperature reading by the thermocouple, indicating a higher level of energy transmission.

CONCLUSIONS

This work shows that absorption is very sensitive to the level of carbon black in the polymer formulation and provides some data to quantify this effect. This sensitivity occurs at very low levels of carbon black. Thus, when a polymer is selected for infrared welding, it will be important to know the concentration of carbon black in the formulation. If it falls below 0.07 percent, there will be increasing depth of heating and less surface heating.

Levels in excess of 0.03 percent carbon will heat primarily by surface absorption of the infrared radiation. In this latter case, the creation of a significant depth of melting will depend on the relatively slow process of conduction. However, if changeover times are short, this latter method will approach high temperature hot plate welding where surface decomposition is tolerated so long as the decomposed material is squeezed out as flash. One significant risk is that the joints may not be as strong as when a deeper melt zone is created.

Because infrared welding has the ability to penetrate polymers and heat them, it offers the potential for stronger joints because a deeper melt zone is created by absorption, at once, rather than by conduction through the polymer.

Plastic parts with a thin black layer on one side can be continuously welded in place. For example, if the black layer is one part of a bilayer, coextruded sheet, it could be unrolled and welded in place. The heat needed for welding would be generated precisely where it is needed, minimizing damage to the part and allowing the joining of thin polymer films. Thin films, particularly when coated on a metal, that are hard to join by other methods should be readily and rapidly joined by infrared welding.

Polymers of different colors can be expected to weld differently by infrared welding. Not only are the issues of pigment-polymer interactions present such as the differences in weldability caused by white (titanium dioxide), black (carbon black) or other pigments, but heating times and depth of heating are likely to be affected by part color. This kind of phenomenon can already occur in conventional hot plate welding when, in some cases, red and black parts weld differently. It can be expected to become even more of an issue with infrared welding. These effects can be easily handled, but workers must be aware of their presence and how to control them.

REFERENCES

1   R. A. Grimm, Through-Transmission Infrared Welding of Polymers, Conference Proceedings of the SPE ANTEC, 1996, Indianapolis, IN., p. 1238.

2   Y. S. Chen, A. Benatar, Infrared Welding of Polybutylene Terephthalate, Conference Proceedings of the SPE ANTEC, 1995, Boston, MA., p. 1248.

3   M. Ference, H. Lemon, R. Stephenson, **Analytical and Experimental Physics**, Second Edition, *The University of Chicago Press,* Chicago, IL, 1956, p. 550, and similar physics or optics textbooks.

Laser Transmission Welding of Thermoplastics: Analysis of the Heating Phase

H. Potente, J. Korte, F. Becker
University of Paderborn, Institute of Polymer Engineering, Pohlweg 47-49, 33098 Paderborn, Germany

ABSTRACT

In laser transmission welding, the parts to be joined are brought into contact prior to welding, and the heating and joining phases take place simultaneously. The laser beam of the Nd:YAG laser penetrates the transparent part being joined and is converted into heat by the absorbing part. The transparent part is similarly heated and plasticized by means of heat conduction, thereby ensuring that the parts are welded together.

When the heating phase was analyzed, it was seen that if the part that absorbs the laser beam has a high absorption constant, this process phase can be readily described by a physico-mathematical model, by analogy to single-sided heat impulse welding. A comparison of calculated and measured melt layer thicknesses showed that, by introducing a correction factor, it is possible for this model to be successfully used for the case of a low absorption constant as well.

PROCESS SEQUENCE FOR LASER TRANSMISSION WELDING

The principle behind the laser transmission welding process is shown in Figure 1. If this process is to be applied, then one of the semi-finished products must be transparent to the laser beam and the other material must have a high absorption constant.

Since the Nd:YAG laser beam penetrates a long way into the majority of plastics which do not contain additives, this condition imposed on the semi-finished products can be fulfilled through dissimilar pigmentation of the material. A plastic which looks colored to the human eye is still perfectly capable of being transparent to an Nd:YAG laser beam.[1]

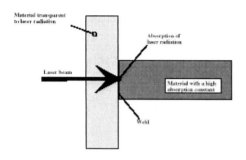

Figure 1. Principle behind laser transmission welding.

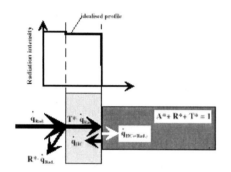

Figure 2. Energy situation in laser transmission welding.

This combination of materials ensures that the laser beam is transmitted through the first semi-finished product almost without obstruction and is then fully absorbed by the second semi-finished product (high absorption constant) in the layers close to its surface (Figure 2).

The heat created in this way is transported by thermal conduction into deeper layers of the absorbing part and also into the part that is transparent to the laser beam. The absorption produces a temperature increase in the joining zone. The plastic melts, improving the thermal contact between the parts being joined and - in the case of parts that were in contact with each other without being subject to pressure at the start of welding - creating an internal joining pressure on account of the volumetric expansion.[2] As with film welding, the heating and joining phases take place at one and the same time in this laser welding process too.

ANALYTICAL DESCRIPTION OF THE HEATING PHASE

During the heating phase, the laser beam penetrates the first part to be joined and is absorbed by the second part. If the absorbing part contains a sufficient quantity of absorbing additives, then it can be assumed, given its high absorption constant, that the radiation energy is completely absorbed in a very thin layer (surface absorption). The remaining transport of heat into the two parts to be joined takes place through heat conduction. Figure 3 shows the melt layer thickness in the transparent part expressed in terms of the melt layer thickness of the absorbing part as a function of the carbon black content of the absorbing part. If the absorbing part contains a small quantity of carbon black, then the radiated energy will be absorbed over a broad layer of material. A relatively shallow temperature profile then results, with a slowly-increasing contact surface temperature. Since the heat conduction in the transparent part is a function of the temperature gradient, only a small part of the radiated energy introduced reaches the transparent part. With a high carbon black content, by contrast, the absorbing layer is very thin, and the energy applied is conducted equally into both parts.

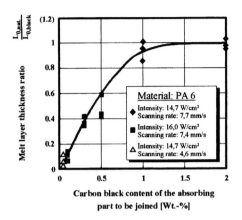

Figure 3. Melt layer thickness ratio as a function of the carbon black content of the absorbing part.

As can be seen from the experimental values in Figure 3, if the carbon black content of the absorbing part is sufficiently high, the thickness of the melt layer will be identical in the transparent and absorbing parts. This means that, assuming a high absorption constant or a high carbon black content in the absorbing part, an identical temperature distribution can be expected in both parts being joined. Exploiting this symmetry, it is possible to observe each part in isolation when analyzing the heating phase. With the assumptions set out below, it can also be assumed that the physico-mathematical model presented[3] to describe the temperature progression in single-sided heat impulse welding is transferable to laser transmission welding in cases where the absorbing part has a high absorption constant:

a) $T(t = 0) = T_0 = T_U$ (temperature of part to be joined at $t = 0$ equivalent to ambient temperature)

b)   no outflow of melt from the joining zone

c)   ideal material behavior

d)   convective heat transfer negligibly small

e)   heat conduction along weld seam and seam width negligibly small

f)   constant heat flow, $\dot{q}_0$

g)   the parts to be joined correspond to a semi-infinite space

h)   both parts to be joined display the same thermal conductivity

From the general differential equation for heat conduction

$$\rho c_p \frac{\partial T}{\partial t} = \Delta(\lambda_H T) + \dot{\Phi} \qquad [1]$$

where: t - time, ρ - density, c_p - specific heat capacity, $\dot{\Phi}$ - internal heat source, λ_H - thermal conductivity. It then follows for laser transmission welding by analogy to single-sided heat impulse welding that:

$$\frac{\lambda_H (T - T_U)}{2\dot{q}\sqrt{at}} = ierfc \frac{z}{2\sqrt{at}} \qquad [2]$$

with the assumption:

$$\dot{q}_0 = 0.5\dot{q}_{total} \qquad \qquad [3]$$

where a - thermal diffusivity, z material depth.

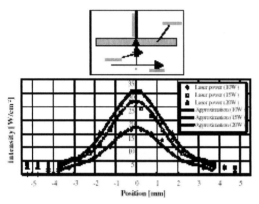

Figure 4. Distribution of the radiation intensity behind a part in natural-color PA 6 (thickness: 5 mm).

For purposes of the experimental verification of equation [2], the distribution of the radiation intensity behind one of the parts to be joined (thickness = 5 mm), in natural-color PA 6, was first measured with different laser powers (Figure 4). The laser beam is widened as a result of the incident laser transmission being scattered by the spherulites in the parts to be joined. This then gives the broad intensity distribution shown in Figure 4. The intensity distributions established here tally in qualitative terms with the measurements presented elsewhere.[4]

To permit their further use, the measured intensity distributions are approximated by a polynomial of the fourth order, and the intensity averaged out over a representative beam diameter d'_s of 7 mm. The heating time at a random point on the surface being joined is then:

$$t_E = d'_s / v_{scan} \qquad \qquad [4]$$

where v_{scan} - scanning rate.

Equation [2] was derived on the assumption of an absorbing part that had a very high absorption constant, which results in a uniform introduction of energy into both the parts to be joined. When the absorbing part has a low absorption constant, equation [2] has a correction factor K* added to it in order to allow for the different temperature profiles that prevail in the parts being joined.

$$\frac{\lambda_H (T - T_U)}{2q_0 \sqrt{a_{eff} t_E}} = ierfc \frac{K^* z}{2\sqrt{a_{eff} t_E}} \qquad \qquad [5]$$

where: $K^* = 1$ (transparent part to be joined) and $K^* < 1$ (absorbent part to be joined), and

$$a_{eff} = k_1 + k_2 \ln(t_E)$$ [6]

where: k_1, k_2 - correction factors.

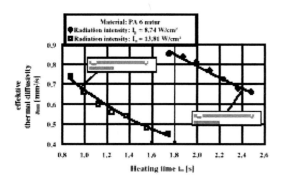

Figure 5. Approximated profiles of the effective thermal diffusivity.

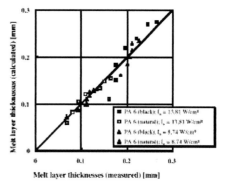

Figure 6. Melt layer thicknesses, measured and calculated.

Equation [5] shows the approximated profiles of effective thermal diffusivity for a part in natural-color PA6, 5 mm thick, with different radiation intensities acting on the joining surface. In the case investigated here, the carbon black content of the absorbing part was 0.5 wt%. The correlation between thermal diffusivity and radiation intensity is due to the fact that the intensity distribution has been averaged out and a corresponding assumption made in respect to the heating time.

The joining surface temperature $T(z=0)$ is identical in the transparent and the absorbing part. In consideration of equation [5], the following coupling condition results

$$\dot{q}_{0,nat} \sqrt{a_{eff,nat}} = \dot{q}_{0,black} \sqrt{a_{eff,black}}$$ [7]

Figure 6 shows the melt layer thicknesses calculated with the aid of equation [5] in the absorbing and the transparent part, plotted over the experimental values. There is seen to be a good level of agreement.

REFERENCES

1   H. Potente, E. Pecha, J. Korte, Entwicklungstendenzen beim Laserschweißen von Kunststoffen, *Plastverarbeiter*, **46** (1995) 10, 58-64.
2   M. Welz, H. Pütz, Schweißen von Thermoplasten mit Laserstrahlen, DVS Plenartagung, Würzburg, (1995).

3   H. Potente, Fügen von Kunststoffen, Vorlesungsmanuskript, Universität-GH Paderborn, 1990.
4   R. Klein, Bearbeitung von Polymerwerkstoffen mit infraroter Laserstrahlung, Dissertation, RWTH Aachen (1990).

Interaction of Lasers with Plastics and Other Materials

Shiv C. Dass

Lasertechnics, Inc., 5500 Wilshire Avenue, Albuquerque, NM 87113, USA

INTRODUCTION

The industrial application of lasers has been growing for more than the past twenty years. Today, laser technology is used in all walks of life. Just to name a few, the lasers are used in medical procedures, surgery, cutting, drilling, welding, soldering, alignment, product marking and coding, etc. Product marking and coding is of special interest to us at Lasertechnics. This field is relatively new and therefore, it is little known in the industry in general. However, laser coding has been receiving wide spread acceptance in packaging goods and growing at a fast pace.

Traditionally, the most common method of coding and marking packaged products is to use ink stamping or ink jet printing. These methods make extensive use of inks, solvents and other chemicals. Although the initial cost of ink coding systems is relatively low and they appear to be simple and flexible, ink methods present problems. Many users are becoming increasingly frustrated with ink markers because of clogging ink jets, low reliability, smudged and/or illegible codes, high maintenance, cost of ink and solvents and their disposal problems. Recently, the use of some solvents is completely banned in some places by the environmental agencies because of toxicity.

The new technology of laser coding, on the other hand, offers significant advantages over ink coders. The ink systems are being rapidly replaced with laser coding systems. This article is intended to provide a brief introduction to lasers and their application in laser coding. The successful applications depend on how the laser light interacts with different materials. Therefore, the type of material and type of lasers become important.

LASER DESIGN

The name LASER is an acronym of Light Amplification by Stimulated Emission of Radiation. A laser is a device that produces an intense, concentrated, and highly parallel beam of light. There are three basic ingredients required for the laser operation. They are as follows:

1 Lasing medium
2 Pumping source, such as a flash lamp or high voltage discharge
3 Reflecting mirrors at both ends of the medium

The atoms or molecules are excited to a higher state of energy by supplying an intense light or electrical energy. These atoms or molecules fall back to the ground state by giving out light energy (photons). These photons stimulate other atoms or molecules to produce more photons. More and more photons are produced when they surge back and forth between the two mirrors. This becomes self-sustaining and the system will oscillate, or lase, spontaneously (Figure 1).

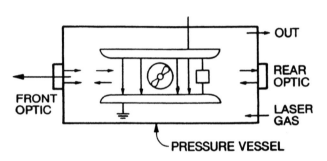

Figure 1. Laser design concept.

A lasing medium is necessary in order to create a lasing action. Based on the medium used, there are three types of lasers (Table 1).

LASER MARKING AND CODING

Almost all packaged goods such as food, beverages, pharmaceutical, and other consumer items require on-line coding with date, lot number, barcode, etc.

Lasers have become the preferred coding method in the packaging industry. Today's industrial lasers provide high quality, sharp, permanent, clean, highly reliable marks, and are free from toxic inks and solvents. In addition, the running costs are extremely low as no ink or solvents are required. Based upon the cost of consumable items alone, a laser marking system can pay for itself within the first eighteen to twenty-four months of operation when compared to ink jet coders. The code changing flexibility is also available.

By far, the most common method of coding packaged products is by using a CO_2 laser at 10.6 μm wavelength. This is because CO_2 lasers can code with a very high speed on the moving object. Also, they are highly reliable, easy to use, and of a relatively lower cost.

Table 1. Laser types

Laser type	Wavelength, μm	Application
Gas lasers		
He Ne, visible	0.632	Scientific, printers, pointers, measuring instruments
N$_2$, UV	0.337	Scientific, cutting
CO$_2$, IR	10.6	Marking, cutting, drilling, welding
Excimer, UV	0.193-0.351	Marking, eye surgery, photochemistry, lithography
Solid state lasers		
Nd:YAG, IR	1.064	Marking, cutting, welding
Alexandrite, near IR	0.7-0.8	Drilling, cutting
Ti-sapphire, near IR	0.812	Medical
Ruby, visible	0.694	Scientific
Semi-conductor lasers		
Ga Al As, IR	0.850	Printers, communication
In Ga Al P, visible	0.670	Printers, alignment
Ga As, IR	0.904	Printers

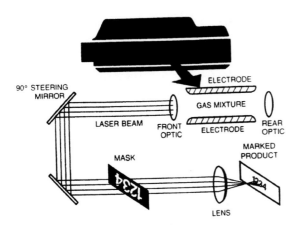

Figure 2. Product coding with CO$_2$ TEA laser.

The pulse CO$_2$ TEA (Transversely Excited Atmospheric pressure) lasers have been coding products for the last fifteen years. The TEA lasers use a stencil or mask, into which the code to be marked has been etched. The laser fires a pulse of very intense light through the mask. An image is then projected onto the product by using special lens(es). The beam can be intensified several fold onto the product to achieve good contrast. A pulse of 5 Joules can typically mark an area of 200 to 250 square mm. The code can be changed by simply changing the mask manually or using a fully programmable device (Figure 2).

Now a new, advanced technique of dot matrix laser coding is also available. This method also uses CO$_2$ laser pulsing with a very high speed, producing a series of dots which are deflected under computer control onto the moving products. The characters are formed in the same manner as in an inkjet, but without ink. Since there is no mask involved, this system provides full flexibility of changing the code.

A small percentage of coding and marking applications are done with Nd:YAG laser at a wavelength of 1.06 μm and Excimer lasers at 0.308 μm.

MATERIALS AND THEIR INTERACTION WITH LASERS

The packaging materials can be classified into the four categories as follows:
1 Paper products
2 Metals
3 Glass and ceramics
4 Plastics such as PE (also HDPE and LDPE), PP, PS, PET, PVC, ABS, etc.

PAPER PRODUCTS

A major criteria in producing a laser code is to produce a high contrast in the largest possible area with the minimum amount of laser energy. This can easily be achieved on paper products such as labels, bags, cartons, etc., if an inked target area is available on the paper. When the laser beam strikes on the ink, it acts like a heat pulse, and therefore instantly vaporizes the ink off the paper, exposing the color under the ink. The same phenomenon takes place on inked, painted, or anodized metal and plastic.

A micro-thin layer of the ink is all that is necessary to achieve good contrast. Most packaging material manufacturers provide such a target on the product to be coded with the laser. This ink target is preprinted along the rest of the printing on the label, and the laser code is applied to the target area during the production process.

METALS

Marking on bare metals is achieved by laser engraving with Nd:YAG laser. The pulse or CW CO_2 lasers described above cannot mark on bare metal because the beam is reflected off the surface without any absorption. Therefore, metals can only be marked if they are anodized, painted, or coated with an absorbing material.

The thickness of the ink or paint on the metal is important. Not enough beam is absorbed if the coating is too thin. For example, on most beverage cans, the ink is so thin that an acceptable contrast is not achieved. On the other hand, aluminum foil pouches used in pharmaceutical and food packaging are coated with paper and/or ink to produce good contrast when coded with the laser.

GLASS AND CERAMICS

The glass and ceramic types of materials easily produce marks with the CO_2 laser without any special coating of ink or paint. As the laser beam hits the glass surface, it only gets absorbed

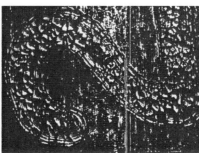

Figure 3. Melted glass film before breaking (left); after breaking the glass film into microcracks (right).

within the depth of a few microns. The surface temperature reaches anywhere from 1000°C to 2000°C creating a soap bubble-like envelope on the surface of the glass. As the glass cools off, these envelopes break into micro-cracks under mechanical stress. This creates a "frosty" looking mark. Since this is a localized and surface phenomenon only, the integrity and strength of glass are not compromised. Figure 3 shows the magnified glass mark. The micro cracks can be easily noticed in the photograph.

The Nd:YAG lasers are unable to interact with glass because the glass is transparent to the Nd:YAG laser beam at the wavelength of 1.06 μm. Therefore, no mark is possible with this beam. However, Nd:YAG and Excimer do interact with ceramics, producing good contrasting marks. Ceramic, ICS, and ceramic chip capacitors are usually marked with these types of lasers.

PLASTIC

Laser technology is growing rapidly in the plastic coding area. A number of plastics are ideally suited for laser marking. However, most plastics require a higher concentration of the laser beam, resulting in a smaller mark area. The interaction of the laser beam with plastics varies.

The pulse CO_2 laser produces a nice, gold color contrast on PVC. Most plastics absorb the CO_2 beam. Some of the plastics partially absorb and partially transmit, while others such as PET are almost completely transparent at 10.6 μm. PET absorbs the CO_2 beam and produces good, legible marks by melting the plastic, although there is no color change.

When mixed with special pigments, most plastics interact with the laser beam to produce a good color contrast. Additives such as mica powder, titanium dioxide, China clay, kaolin, talcs, calcium carbonates, titanium dioxide coated mica (AFFLAIR), etc. can produce the desired results with the laser.

When the laser pulse hits the plastic material with pigment, the pigment particles absorb the beam very heavily. This increases the temperature of the pigment particles so high that the partial carbonization of the plastic takes place in the vicinity of the pigment. Thus, a high contrast, dark gray mark with sharp edges is achieved in polyethylene and polypropylene type plastics. A high peak power of the order of twenty to twenty-five mega watts is required to achieve high contrast. The TEA CO_2 laser with a pulse length of one hundred nanoseconds (total pulse length one to two micro-seconds) can easily achieve such a high peak power. The same results are not achieved if the peak power is not as high.

Material thickness also plays an important role in providing high contrast. If the plastic is one millimeter thick or more, then a small percentage (0.2% to 1 %) of the pigment is enough for a good contrast. If the base material is very thin, on the order of submillimeter or a few microns, a larger percentage - as high as 10% - may be necessary to produce good contrast.

The Excimer lasers with the wavelength in the UV range (0.103 to 0.351 μm) can also be used for marking on plastics and pigmented plastics. The mechanism that changes the color, is photochemical in nature, rather than carbonization.

SUMMARY

Laser technology in product coding has come a long way in the past fifteen years. More and more ink markers are being replaced with laser markers. The advantages are: mark permanence; virtually maintenance free; no consumables like ink and solvents; and high quality appearance of the code. The availability of pigments overcomes some of the laser limitations. The CO_2 lasers are the most commonly used in coding. A small percentage of coding applications use Nd:YAG and Excimer lasers.

Future laser systems will produce higher energy, higher speeds, larger marks, and a compact size. The systems will be simple and highly flexible to change the code under computer controls.

Customized Decorating of Plastic Parts with Gray Scale and Multi-Color Images Using Lasers

Terry J. McKee, Landy Toth
Lumonics Inc.
Wolfgang Sauerer
BASF AG, Germany

ABSTRACT

Customized decorating with sophisticated graphic images applied using lasers is a new value-adding process applicable to plastic parts. Many of the techniques familiar for printing on paper such as halftone and dithering are applicable to polymers using indelible direct-write laser marking technology. The authors will describe computer-software methods applicable to transfer graphics to a format suitable for laser marking and provide examples of decoration applied to suitable polymers, including color marking

INTRODUCTION

Using a laser to directly place functional codes on products (date codes, batch codes, bar codes) is one of the most common industrial applications of lasers, with an estimated installed base exceeding 20,000 units. When compared to other coding solutions, direct marking with a laser provides an indelible mark using a non-contact, fast, and reliable process. As the ability of lasers to directly mark suitable polymers has become evident, a new and potentially more lucrative application for lasers has materialized: decorative marks on products in the same market areas. In most of the same industries which require functional coding, attractively-designed packaging is a very important factor in product identification and market share. The best product in an unattractive package risks anonymity. The design and manufacture of attractively-decorated packaging accounts for a significant portion of product cost not only in consumer packaging but in electronics, automotive, and many other industries. Because of the advantages already apparent in laser coding, the use of lasers for decorating, and particularly in customized decorating, of polymers has recently gained considerable interest.

Laser coding is typically an on-line process taking place as a bottle is filled or an electronic part is encapsulated. Decoration can often be an off-line process where speed is not an important factor and low cost solutions are effective. However, for the ultimate package, laser decorating can offer a distinctive mark not achievable by any other way: a color-change indelible graphic image of a leaf marked into an otherwise-standard polyethylene skin cream container; a customized logo or graphic image applied to an ABS electronic part before or after assembly of a personal digital assistant, PDA; a facial image marked into a white polyethylene plastic credit card. Expensive perhaps, but in many markets the distinctive decorating solution buys market share?

The most sophisticated laser marking technology which is most suitable for decorative marking of graphic images is the directed beam "laser-writing" scenario, usually utilizing the solid state Nd:YAG laser. In this scheme, a PC computer drives two galvanometer-mounted mirrors which scan the YAG laser beam in a suitable pattern focused on the sample to be marked. As opposed to the common printer which raster prints a bitmap matrix of dots on the page, a laser writer marks a vector image composed of lines, closed polygons and fill patterns usually associated with "writing". Because the computer is the brains for the laser-writing technique, off-line input and manipulation of graphical images is feasible and sophisticated decorative laser marking becomes possible. Other lasers than the YAG laser can also be used within the laser-writing format if that is what is required to mark the polymer in question. A discussion of the various laser marking techniques available to mark polymers can be found in references.[1,2]

This paper will discuss the techniques whereby sophisticated graphic images can be manipulated to optimize their appearance in decorative laser marking of polymers. This will involve the manipulation of digitized bitmap images to obtain vector images suitable for laser-writing, the choice of fill patterns to achieve the desired gray scale image quality, and evaluating the interaction of the laser with the desired polymers. Various examples of grayscale and color decorative images marked on suitable polymers will be given.

DIRECTED-BEAM LASER WRITING

A schematic representation of the directed-beam laser writing scheme is shown in Figure 1. Although a carbon dioxide or excimer gas laser can be used, the most common laser used for directed-beam writing is the solid state YAG laser operated in the cw Q-switched mode. With typical laser powers of about 20 watts average power and a pulse repetition rate of 20,000 pulses per second, the laser beam when focused on the sample makes a small (about 0.1 mm diameter) dot for each pulse. By directing the laser beam using fast computer-driven mirrors

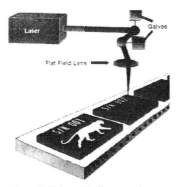

Figure 1. Schematic diagram of directed-beam laser writer.

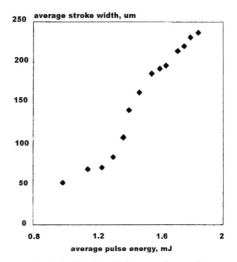

Figure 2. Marked white line thickness vs. laser pulse energy.

mounted on galvanometers, marks can be made directly into the surface of the polymer, most often *via* a thermal process but also sometimes *via* a photochemical process. The size of each dot marked on the surface is determined not only by the energy per pulse of the YAG laser, but also by the "degree of markability" of the polymers - absorption coefficient, thermal conductivity, melting temperature, and susceptibility to local pyrolysis or color change. The choice of the best laser wavelength and the use of suitable polymers and additives are the two most important tools to achieve a good laser mark.[3-6]

For directed-beam YAG laser marking, each laser pulse corresponds to a dot marked on the sample. Lines are formed by a series of overlapping dots. Depending on the markability of the material, the width of the line varies depending on the laser pulse energy and the dot overlap. In Figure 2, the width of a white line marked with the YAG laser using 833 dots per inch is shown as a function of laser pulse energy for red acetal copolymer POM (Ultraform® BASF). This curve is analogous to the dynamic response function used for thermal print media.

An image written by the laser writer is a collection of objects: lines, connected polygons, and fill patterns - sometimes as many as 500 objects based on the vector file format. The vector file format is considerably simpler and the laser writing speed is considerably faster when compared to the billions of pixels required for the bitmap or raster file format used for continuous tone gray scale images. Common vector file formats are *.dxf (AutoCAD), *.cdr (CorelDraw), *.ttf (Truetype fonts), *.afm (Adobe type 1 font), *.cgm, and *.met. Gray scale can be replicated in a non-continuous manner in vector graphics by using different forms and densities of fill patterns, some of which are shown in Figure 3. By combining lines, polygons and fill patterns in a vector image, a close replica of the continuous tone gray scale image can

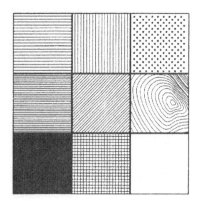

Figure 4. Continuous tone bitmap image printed on paper (left), and vector image laser marked on white polyethylene card (right).

Figure 3. Various fill patterns for vector images.

be made as shown in Figure 4. In Figure 4, the image on the left is a continuous tone image with 16 levels of gray scale printed on paper using a 300 dpi inkjet printer, and the image on the right is a vector image laser marked on a white HDPE card containing a mica-based additive (0.3% LS 820, EM Industries).

IMAGE MANIPULATION

Although various types of commercial software can be used to manipulate images, we commonly use the CorelDraw 6 software package. Graphic files received in any of the common vector file formats can be imported directly into CorelDraw 6. However, when a scanner, digital camera, Kodak photoCD, or other technique is used to digitize an image, the first step must be to transform the bitmap graphic format to the smaller-file-size vector format suitable for the laser writer. To change the *.bmp or *.tif bitmap file to a vector file, the OCR Trace software in the Corel 6 package can be used either in fill pattern mode or sketch mode. Once the vector image has been generated, it can be imported into CorelDraw 6 and that software can be used to touch up lines and simplify the image if necessary, to evaluate levels and patterns of gray scale, and to export the modified graphic image to the laser writer. Software manipulation of a complex image in CorelDraw 6 usually involves breaking the image into objects on different layers, trimming objects in the background to remove overlapping segments, choosing fill patterns for each object, and recombining layers to form the final image. The goal is to simplify the file as much as possible while still retaining the visual quality of the image.

WRITING GRAYSCALE AND COLOR IMAGES ON POLYMERS

Many polymers mark well using the YAG laser; for others, additives such as mica, titanium dioxide or pigments can be used to improve the mark contrast significantly.[1-6] A typical mas-

Figure 5. Vector image YAG laser marked on black polyamide showing various line thicknesses and fill patterns.

ter batch will contain many additives some of which may also affect laser markability; therefore, it is best to first evaluate the markability of the sample in question with a variety of lasers. A good mark can then be improved by optimizing laser parameters such as wavelength, peak power, dot density and dot overlap. In some cases where low absorption of light by the polymer is a problem, it may then be necessary to resort to additives to get an aesthetically-attractive mark.

It is important to be able to change the width of a marked line by changing laser power. This means that the dynamic response function of the polymer (cf. Figure 2) should exhibit a monotonically-increasing region over a reasonable dynamic range of laser pulse energies. This facilitates the marking of thin or bold lines simply by changing the laser pulse energy as controlled by the laser writer software. On the other hand, a long flat region in the sensitivity function may indicate a polymer with low absorption for which the marked line is an ever-deeper groove with unattractive visual appearance. A vector grayscale image of a car marked on black nylon containing a mica-based additive (1% LS 830, EM Industries) for which various line thicknesses and fill patterns are used is shown in Figure 5. This is an example of a monotonic grayscale image, that is the laser makes single color (white) marks on the black polymer primarily by "foaming" on the surface.[7]

In some cases, depending on the pigments, additives, polymer and laser used, it is possible to generate two different colored dots (white or black) depending on the pulse energy of the laser. In this case, lines and fill patterns can be made either black or white, or various shades of gray in between using different combinations and fill densities of dots. This greatly facilitates the use of gray scale in vector images. Examples of decorative laser marks on polymers which allow both white and black marks are shown for blue PBT (BASF Ultradur® B4520) using a green (frequency-doubled YAG) laser in Figure 6, and for yellow acetal copolymer (BASF Ultraform® N2320) using a diode-pumped YAG laser in Figure 7.

Figure 6. Green laser marks both black and white on blue PBT.

Figure 7. Diode-pumped laser marks black and white on yellow acetal copolymer (POM).

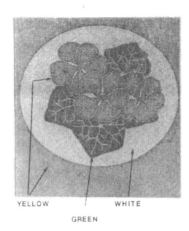

Figure 8. Yellow flower with green leaves and white background marked on yellow polyamide using green laser.

It has also been predicted that color marking of plastics is possible, that is the laser induces a color change of the polymer to a color other than white or black. In Figure 8, we show a decorative mark achieved using the green (frequency-doubled YAG) laser to mark yellow polyamide 66 (BASF Ultramid® A3K) to form a colored image of a yellow flower with green leaves and a white background. To our knowledge, this is the first demonstrated example of a true laser-marked multi-color image in plastic.

ACKNOWLEDGMENTS

The authors would like to thank Jeff Babich (EM Industries) for providing polyethylene and nylon samples containing mica-based additives.

REFERENCES

1   T. McKee, Proc. RETEC. (Div. Decorating and Assembly), Society of Plastics Engineers, Chicago, Sept 21-22, 1994, 217-230.
2   T. McKee, Laser Marking, *Plastics Formulating & Compounding (US)*, Nov 1995, 27-32; T. McKee, How Lasers Mark, *Electrotechnology (UK)*, **7** no. 2, April 1996, 27-31.
3   T. Kilp, Proc. ANTEC 91, Society of Plastics Engineers, San Francisco (1991), pp. 1901-1903.
4   T. McKee, Packaging Technology & Engineering, August 1995, 48-52.
5   L. Y. Lee, D. E. Roberts, B. L. Vest, K. Tonyali, Laser Printable Black Cable Jacketing Compounds, Proc. International Wire & Cable Symp., 1995, 823-828.
6   G. Graff, Additives make possible laser marking of polyolefin components, *Modern Plastics*, April 1996, 24-25.
7   C. Herkt-Maetzky, W. Robers, Investigation of Color Change Reaction During Laser Marking of Plastics, *Macromol. Symp.*, **100**, 1995, 57-63.

Color Laser Marking: A New Marking and Decorating Alternative for Olefins

Alan Burgess, Ke Feng
M.A. Hanna Color, M.A. Hanna Company

ABSTRACT

A new technology to color mark molded polyolefin parts with lasers will be described. Through use of custom additive packages rather than base-resin manipulation, this technology offers dramatic improvements in the ability to mark and decorate olefins and a broad range of other thermoplastics using a more extensive color palette than previous laser-marking technology.

Benefits of the new second-generation technology will be compared and contrasted with first-generation laser marking and with direct printing and use of labels - traditional methods for marking and decorating plastic parts. This paper also will discuss olefin application areas where the technology is expected to penetrate.

PROBLEMS WITH DECORATING & MARKING PLASTIC PARTS

Decorating, labeling, marking, or coding plastic parts is generally a challenge for design engineers. Direct printing, the standard marking method, can be accomplished *via* ink pad or ink jet printing, ink filling, sublimation printing, embossing, or hot stamping. However, information that is printed on plastic is rarely permanent due to the non-porous, often chemically resistant surface of these materials. In fact, owing to their excellent chemical resistance, olefins are notoriously hard to print on without the use of surface treatments. Worse, in certain environments, this surface mark can have poor scratch, wear, and solvent resistance. In many markets where durability of the mark is of special concern, another step (overcoating) is required after printing to protect the mark. Direct printing also has limitations in terms of image structure and density, the amount of information that can be embedded in the mark, and the flatness of the surface that can effectively be printed.

Use of self-adhesive labels is another option, but this approach poses similar problems. Labeling operations can have high scrap rates (up to 30% in some operations) due to off-center placement of labels, which may slip before the adhesive is fully set, or which may tear or be scuffed during handling.

Liability issues can arise with both direct printing and labeling when warning signs or labels wear beyond readability or simply fall off a product.

Also of concern are the unfavorable environmental implications for labeling and printing operations. As previously mentioned, the high scrap rates characteristic of labeling operations generate potential regrind. However, because paper does not melt, scrapped parts with labels cannot be melt reprocessed without first removing the label, an often cumbersome and costly operation. Printing on plastics involves the application of aqueous- or alcohol-based inks, often in conjunction with solvent-based primer coats or corona or flame treatments, plus topcoat protective coatings. These chemicals require additional storage, some require special handling, and additional energy is consumed to apply them to the part. Printing can also generate significant scrap, depending on the complexity of the surface being printed, the criticality of placement of the printed material on the part, and the durability required of the mark.

Finally, direct printing and labeling are relatively expensive, multi-step secondary operations, usually performed off-line from the initial molding process. The more steps in a production process, the greater the potential for error, not to mention the additional time, floor space, and costs of the operation.

FIRST-GENERATION LASER-MARKING TECHNOLOGY

An alternative technology for marking olefins and other plastics, one that is several years old, is the use of commercial industrial lasers to char or etch a permanent mark into the wallstock of a polymeric part. While many plastics are normally transparent to laser energy, modification of the base resin *via* additives and fillers, often combined with an increase in beam intensity, improves a polymer's absorption of laser energy. This manipulation causes local pyrolysis color change in the part's wallstock upon exposure to the laser energy, resulting in a permanent mark that will not rub, scuff, or scratch off, and that can be placed on a more complex, curved surface. Marks can be as simple as a bar code, or as complex as a logo or the 2D codes now favored by companies like UPS. A further advantage is that the technology can be applied in-line or near the molding operation, reducing handling and floor-space requirements. Unlike printing, no additional solvents, inks, or chemicals are involved that require special handling; no primer coats or surface treatments are required to prepare the part for marking; and no protective topcoat is needed to protect the mark from scuffing or damage. Unlike labeling operations, scrapped laser-marked parts can be re-ground and melt reprocessed without concern of melt-stream contamination. The durability of the mark, the high

Table 1. Advantages of laser marking vs. other marking technologies

Marking process	Speed	Durability	Image flexibility	Contrast
Laser marking	Good	Good	Good	Good
Chemical etch	Good	Good	Poor	Poor
Photo etch	Good	Good	Poor	-
Ink jet	Good	Poor	Good	Moderate
Mechanical stamping	Good	Good	Poor	Poor
Nameplates	N/N	Moderate	Poor	Good
Casting/molding	Good	Good	Poor	Poor
Pneumatic pin	Moderate	Good	Moderate	Poor
Vibratory pencil	Poor	Good	Good	Poor
CO_2 mask marker	Good	Moderate	Poor	Moderate

speed of the marking process, and the ability to change the marked image quickly *via* computer are features that set laser marking apart from other marking technologies (Table 1). However, this process also has its limitations.

Since the chemistry of the base resin must be manipulated, first-generation laser-marking technology has been restricted to certain resin families and grades. Further, it generally has yielded a low contrast between mark and background and has been limited by a very narrow color-marking palette. Available colors have generally been black and shades of gray, although more recently some limited shades of colors have been introduced.[1,2,3] Additionally, first-generation technology does not allow clear plastics to be marked without degradation of the polymer. Generally, the limits to the technology also required that the mark always be darker than the background color.

Further, problems can arise because the technology depends on relatively severe (although localized) thermal modification of the material, often deep into the wallstock of the part. This can lead to aesthetic problems such as crack or bubble formation, delustering due to surface alterations from melting, and, in severe cases, de-polymerization.

COLOR LASER MARKING OF PLASTICS: SECOND-GENERATION TECHNOLOGY

Given how promising laser marking is as a decorating and marking medium for plastics, a plastics color compounder initiated a 3-year development program to overcome the limitations of earlier laser-marking technology. Since first-generation laser marking was more an art than a science when it came to formulation technology, the approach taken by the development team in working on the new product was to take a more rigorous, scientific approach. Instead of modifying the base resin chemistry to achieve markability, which necessarily was a slow process and limited the number and types of materials available, the team worked to create formulae that included custom-additive packages and laser beam manipulation parameters. This 2-pronged approach allowed the technology to be applied to a far more diverse group of materials and translated faster and far more cost effectively than the earlier approach requiring materials reformulation.

The result of this research is a new, second-generation technology that allows for color laser marking of a much broader range of plastics for the first time *via* a non-charring method. All the advantages of first-generation laser marking are maintained in this new patent-pending technology. These include creation of an indelible mark; the ability to mark in-line or nearby the manufacturing process; significant systems cost savings vs. traditional, printing-based methods; and a cleaner, more environmentally friendly marking method that is compatible with reuse of in-plant scrap and post-consumer recycling. However, the new technology is also designed to avoid many of the limitations of previous laser-marking systems.

For example, the new technology allows for a far wider range of plastic materials to be marked, using a significantly broader color palette. Currently, all major thermoplastic-resin families can be marked, and work is underway to optimize packages for common thermoset polymers as well. Further, the technique does not destroy the surface of the marked part, which yields more consistent results with better aesthetics than earlier laser marking. Finally, a much broader color palette is available, with far higher contrast between mark and background, including some ranges of color matching.

SPECIAL OPPORTUNITIES FOR COLOR LASER MARKING OLEFINS

Color laser marking is an especially interesting opportunity for manufacturers of olefin parts. One reason the technology is attractive is that olefins, with their excellent chemical resistance and highly polar surfaces, are notoriously difficult to paint, coat, or plate. Hence, primer coats or flame/corona treatments are a must in order to get anything to stick to an olefin substrate. However, additional process steps detract from the otherwise cost-competitive nature of these plastics. And while olefins generally have good wear properties, their relatively soft surfaces are easy to scratch, which can damage labels and printed information, which sit only on the surface of the part. Since laser marking creates a permanent color change of up to 1 mm deep in the wallstock, surface scratching and rough handling will not remove the mark.

So far, clear grades of polypropylene have proven to be easier to color laser mark than polyethylene. Even PP grades with glass and glass/mineral fillers have been successfully marked. Properly formulated additive packages in olefin parts can meet FDA and HB flame-retardant ratings from Underwriters Laboratories. Studies to date indicate that thermal cycling of olefin parts does not affect the permanence of the mark. While it is possible to re-mark a previously laser marked part, it currently is not known whether multiple doses of heavy IR laser energy would affect the light fastness of the mark. Work is currently underway to address this question. To date, a variety of primary colors have been formulated for olefins.

TARGET APPLICATIONS FOR COLOR LASER MARKING OF OLEFINS

The advantages of the new second-generation laser-marking technology are numerous. The range of colors that can be developed on olefins and most other thermoplastic materials is achieved in a fraction of the time and cost required with first-generation systems because it is integrated into an end product's manufacturing process. Second-generation laser-marking technology is well suited to current market trends for increased customization or "customized" products, where unique configurations are produced for each of a large body of customers. Aesthetic applications and vanity labeling - even customized down to the customer's name - are possible. As such, second-generation laser-marking technology is well suited for a broad range of markets. Because the new marking technology minimizes thermal damage to the resin and preserves surface qualities, laser technology can combine highly customized decorative elements with high-visibility locations.

Initial olefin applications will likely focus on the agricultural/industrial, packaging, medical, wire & cable, automotive, large appliance, and housewares/toys markets, but additional uses are expected to develop as the technology proliferates.

For instance, alphanumeric ear tags for cattle and other stock are currently printed. But the mark can fade as it is exposed to weather. Color laser marked tags would offer a cost-effective and permanent solution to this problem. Agricultural storage bins would also be prime candidates for laser marking. Other olefin industrial totes, dunnage, and trays would also be good candidates for laser marking. The permanence of the mark would help ensure that rough handling and general wear and tear did not lead to premature loss of the information.

In packaging applications, caps, closures, and decorated bottles for short-run promotional campaigns could be marked rapidly and cost-effectively. Some packaging companies are already evaluating the system to mark short-run promotional items that require time-sensitive information to promote contests on bottles, caps, and jars. Additionally, manufacturers will now have the ability to control inventory and information to avoid over-stocking. In private labeling operations, a base product can be created and identified by logo or mark at the time of shipment. Private labeling of generic products provides another area for use of the system.

For the medical industry, housings for various durable or disposable medical devices could be marked. The permanence of the mark would help ensure that information stayed put on the device, despite one-time or repeated ETO, E-beam, or cold-chemical sterilization cycles. In the case of disposable devices, the cost-effectiveness of the mark would help hold costs in line while improving aesthetics: Medical needle disposal units are also a target for laser marking, as they have to be both transparent, but also have red identifying hazard warnings. Work is underway to develop a colored mark out of a transparent polymer. Medical equipment manufacturers in particular can benefit from using laser marking to create language-specific, permanent usage and care instructions, without tooling changeovers or secondary operations. Polyethylene medical packaging may also be a candidate for color laser marking.

Another area of interest is the wire & cable industry, where olefin-jacketing materials are now marked in line by inkjet or offset printing before coiling and spooling. Laser marking could replace this marking technology, offering a more permanent, weather- and abrasion-resistant information-carrying system.

In the automotive industry, unpainted, unskinned polypropylene-based interior-trim components such as pillars, IPs, cluster bezels, center consoles, door panel components, handles, switches and buttons, sills, and package shelves are candidates. The ability to add a highly durable, decorative mark to unskinned, unpainted components in economy vehicles, light trucks, and other industrial vehicles has strong economic incentives.

The large appliance market is another probable target, since any place a metallic label is now attached to a plastic component would be a good candidate for laser marking, e.g. on refrigerators, freezers, and air conditioners. Laser marking would produce a permanent mark that would not fall, rust, or scrape off the appliance like metal, foil, and paper labels. There may be even broader opportunities in the small appliance market, which makes use of more olefins. Here, dishwasher-resistant logos and other information could enhance the aesthetic appeal of the units, while improving long-term durability. Similarly, various other housewares could also be laser marked for identification purposes, or to improve aesthetics and provide market differentiation. Toys pose a similar opportunity, and laser marking again offers the benefit of greater durability.

Finally, in the electrical/electronics industry, handheld computer games, calculators, remote controls, and other components would benefit from the permanence of the mark, a reduction in manufacturing and inventory steps, and the ability to quickly change designs without changing tooling or cliches. Even requirements for special colors on function keys or symbols lend themselves to the second-generation laser-marking process. Considerable manufacturing savings also accrue with the ability to replace separate, affixed labels for electrical safety testing standards (UL, TUV, etc.), radio-frequency (FCC, IEC), and serial numbering, all with laser marking.

CONCLUSIONS

By incorporating second-generation, laser-marking formulation technology into the process for manufacturing olefin parts, production flexibility is increased with no aesthetic or part performance degradation. Secondary marking operations can be eliminated. In addition, manufacturers can respond swiftly to short-term changes in the market due to volume or product mix with an increased potential for customized production. Significant savings could be realized from reduced rework and scrap, and part inspection costs; shorter production cycle times; and reduced lead times. Secondary benefits of this technology include reduced floor space requirements, improvements in inventory, better, more permanent aesthetics, and enhanced versatility to custom mark products.

ACKNOWLEDGMENTS

The authors wish to acknowledge the invaluable assistance provided by members of Control Laser Corporation, Lumonics, Lasertechnics, AB Lasers, and E. M. Industries. Without their support, this work would have not been possible.

REFERENCES

1   T. McKee, L. Toth, W. Sauerer, Customized Decorating of Plastic Parts with Gray Scale and Multi-Color Images Using Lasers, Proc. ANTEC, Society of Plastics Engineers, 1997.
2   **U.S. Patent No. 5,576,377**, Nov. 19, 1996.
3   **Japan Kokai Tokyo Koho, Patent No. 96127175**, May 21, 1996.
4   R. Stevenson, Industrial Strength Laser Marking: Turning Photons into Dollars, Excel Control Laser, Inc., 1992.
5   T. McKee, Laser Marking of Polyethylene and Other Polyolefins with Additives, *Plastics Formulating & Compounding*.
6   T. McKee, Laser Marking of Polyethylene, Other Polyolefins Making Waves," *Packaging Technology & Engineering*, Aug. 1995, 47-52; also presented at RETEC Chicago, Sept. 21-22, 1994, Society of Plastics Engineers, Brookfield, CT, 1994.
7   L. Lee, D. Roberts, B. Vest, and K. Tonyali, Laser Printable Black Cable Jacketing Compounds, Proc. 44th International Wire & Cable Symposium, Eatontown, NJ, 1995, pp. 823-828.

Implementation of Beam-Steered Laser Marking of Coated and Uncoated Plastics

Richard L. Stevenson
Excel/Control Laser, Inc.

ABSTRACT

Beam-steered laser marking offers manufacturers a unique combination of speed, performance, non-contact marking and the versatility of computer generated imaging. Although utilized for a wide variety of metallic and non-metallic materials, plastics marking is the most demanding in terms of process development and image composition. High quality images at optimum cycle times requires a working knowledge of beam-steered image generation and the laser's interaction with the plastic.

INTRODUCTION

Beam-steered Nd:YAG (Neodymium: Yttrium Aluminum Garnet) laser marking provides a unique combination of speed, permanence, and image versatility in a non-contact marking process. Laser marking can generate considerable savings in reduced manufacturing costs, tooling costs, elimination of secondary processes, reduced inventory expense, quality control costs, maintenance downtime and elimination of consumables disposal. Laser marking frequently improves the aesthetic appearance of the marking image thereby increasing the products perceived value.

Of all materials, plastics are the most challenging in terms of the lasers' interaction with the material and the required image quality. The wide variety of material chemistries and colors and the aesthetic requirements of most plastics applications require special consideration in both material chemistry and imaging techniques. The successful implementation of laser marking technology requires a working knowledge of the laser markers' function and capabilities and a committed, team approach by the user.

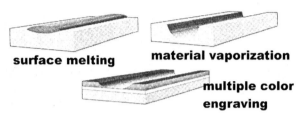

surface melting   material vaporization

multiple color engraving

Figure 1. Surface marking techniques.

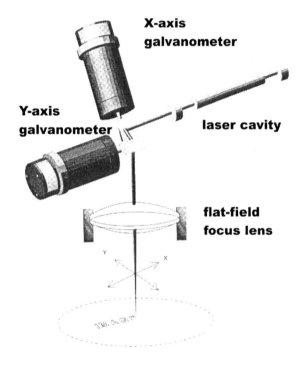

X-axis galvanometer

Y-axis galvanometer

laser cavity

flat-field focus lens

Figure 2. Optical beam delivery system.

MARKING FUNDAMENTALS

Laser marking is a thermal process that employs a high-intensity beam of focused laser light to create a contrasting mark on the material surface. As the target material absorbs the laser light, the surface temperature increases to induce either a color change in the material and/or displace material by vaporization to engrave the surface (Figure 1).

Beam-steered laser marking employs mirrors mounted on high-speed, computer-controlled galvanometers to direct the laser beam across the target surface (Figure 2). Each galvanometer provides one axis of beam motion in the marking field. A multi-element, flat-field lens assembly subsequently focuses the laser light to achieve high power density on the work surface while maintaining the focused spot travel on a flat plane. The laser output is gated to blank the beam between marking strokes.

Marking can be accomplished at speeds of up to 5,000 millimeters per second (197 inches per second) with positioning speeds between marking strokes up to 50,000 millimeters per second (1,968 inches per second). Because the process relies on heat conduction into the plastic, marking speeds are usually slower than the systems maximum capability to allow sufficient conduction to achieve the desired results.

The beam-steered marker can duplicate virtually any black-and-white image including variable line widths and images as small as 0.25 mm (0.010 inch). Present computer imaging

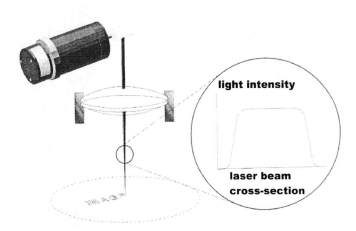

Figure 3. "Top Hat" mode distribution.

technology produces highly intricate graphics with line widths, resolutions, and accuracy well below 0.025 mm (0.001 inch). Because the image is created by "drawing" with the laser beam, the marking time is dependent on the amount and complexity of the text and graphics. With computer generating imaging, any graphic element or the entire marking program can be instantly changed before a new part is positioned for marking.

Nd:YAG lasers amplify light in the near-infrared at 1.06 µm. They are unique among the different types of lasers in that they operate much like an "optical capacitor". In pulsed operation, the Nd:YAG laser stores energy between pulses resulting in peak powers of kilowatts of light energy. A Nd:YAG laser emitting 75 watts of continuous light, pulsed at 1 kHz, emits a train of pulses with peak powers of 110,000 watts. The "optical capacitor' effect provides the peak power necessary to vaporize material. For plastic applications, it is also necessary to run the laser in a "top hat" mode (Figure 3) where the power distribution is fairly even across the cross-section of the laser beam. The top hat mode eliminates "hot spots" in the marking path.

The beam-steered Nd:YAG marker frequently replaces acid and electroetch systems, stamping and punching systems, and those other marking systems that permanently mark products by imprinting or engraving. It also replaces other, less permanent printing systems, including ink jet.

UNCOATED PLASTICS

Most uncoated plastics must be doped with a material reflective to the laser wavelength to prevent over-absorption of the laser light. Too much absorption will result in loss of control of the temperature rise and excessive marking on the surface. Light colored plastics are doped with mica, titanium oxide, or other materials that contain carbon. The heat generated by absorption of the laser light migrates the carbon to the surface producing a contrasting dark mark against the unaltered background plastic (Figure 4).

Plastics are semi-transparent to the near-infrared wavelength of the Nd:YAG laser. Depending on the degree of transparency and the laser output power, the laser beam can alter the

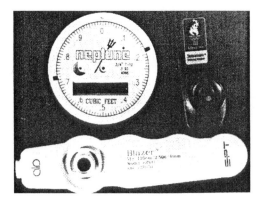

Figure 4. Marking uncoated plastics.

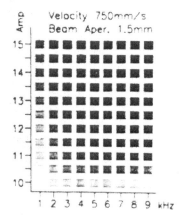

Figure 5. Gray scale values created by altering the laser output power and pulse rate.

material surface to depths of over 0.025 mm (0.001 inch) without achieving vaporization temperature on the surface. If material vaporization occurs, the layer of carbon is thinned and the marking image will appear washed out.

There has been considerable success in altering the depth of carbon migration to create grayscale graphics on light plastics. Adjusting the power and/or pulse rate of the laser controls the depth of penetration and thereby the darkness of the mark (Figure 5). Increasing the laser power will increase the overall depth of penetration and thickness of the carbon layer. Increasing the pulse rate will result in a longer pulse width and lower peak power. The longer exposure also increases the depth of penetration and associated carbon layer.

Dark plastic is doped with a material that produces a lighter color as the material expands and the density decreases. As the temperature of the plastic increases, the plastic expands to form a "blister" on the surface and a lighter colored mark. As with light plastics, the temperature must be tightly controlled to avoid over-absorption. If the temperature rises too high and the blister bursts, material is lost and the mark will lose contrast.

Not all plastics require dopant to achieve a contrasting mark. There are several plastics that yield excellent results without additives. As example, most black polycarbonates produce a snow-white mark without altering the chemistry.

COATED PLASTICS

Coated plastics consist of a solid, translucent or transparent plastic with one or more coats of ink or paint. The marking image is created by achieving vaporization temperature on the surface to remove the top coat and expose the underlying plastic or second coat (Figure 6).

Coated plastics allow a great deal of control over color selection and marking contrast. Transparent plastics allow the designer to use an underlying part to establish the background color (marking image) while the top coat determines the foreground color. Solid plastics es-

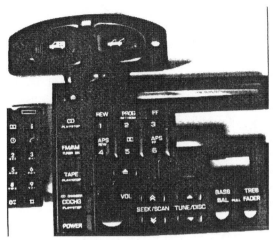

Figure 6. Marking coated plastics.

Figure 7. Top coat removal exposing the underlying white coat on translucent plastic.

tablish their own background with the color of the plastic. Translucent plastics are frequently used for backlit applications. The plastic is initially coated with a white paint and overlaid with a dark top coat. The laser removes the top coat exposing the white paint for day time visibility. When the part is backlit at night, the lighting illuminates the translucent plastic from behind and the marking image appears in the color of the plastic.

The paint or ink used must be conductive to laser processing. Standard paints and inks are not predictable nor controllable when exposed to the laser output. The inks burn easily and can mix the underlying plastic while in the molten liquid state. Laser compatible inks are mixed with a silicone based material reflective to the laser output thereby reducing the inks light absorption and rate of thermal reaction. Paints must be suitable for high temperature processing and be free of any contaminants that may absorb the laser wavelength and speed up the thermal rise.

To achieve a quality image, the top coat must be completely removed with minimal impact on the underlying plastic or secondary coat (Figure 7). To maximize the ratio of light absorption between the two layers, the top coat must always be a dark color and the contrasting underlying layer must be a light color. The dark color will absorb a comparatively higher percentage of the laser light resulting in a higher surface temperature while the light color reflects a higher percentage and minimizes the temperature rise. The underlying plastic, paint or ink should also be thick enough to tolerate a minor amount of material removal during marking.

Marking coated plastics is a multi-step process in which the first marking pass removes the majority of the top coating. The remaining residue is removed with a second, lower power pass to minimize the effect to the underlying material. For precise edge definition, the outline is marked with a heavy edge pass (i.e., 50 kHz, 250 mm/sec, 2.5 watts) followed by a lower power cleanup pass (50 kHz, 250 mm/sec, 1.75 watts). The image is then filled if desired with a heavy fill pass (50 kHz, 650 mm/sec, 6 watts) and subsequent cleanup pass (50 kHz 650 mm/sec, 4.5 watts). Care in determining the process parameters for each pass and the edge and fill beam paths will result in a crisp, high contrast, high quality marking image.

A Note of Caution: Plastics outgas material during laser processing. Some plastics will emit toxic materials requiring special ventilation at the marking area. Do not rely on the laser marker manufacturer to advise you of any potential hazards related to the laser processing of your specific plastic. Laser marking is routinely applied to a wide variety of metallic and non-metallic materials. Although laser manufacturers are proficient on the requirements for processing materials, they are not necessarily knowledgeable on the chemical and physical transitions that may occur.

PREPARATION AND INSTALLATION

Perhaps the most critical element in the successful application of laser marking is the composition of the part programs. When replacing an existing marking technology, you must allow up to six months for conversion of existing artwork to part marking computer programs. Even if the present artwork resides in AutoCAD files, time must be allotted to convert the files to optimized marking programs.

Many users start with thousands of sheets of Mylar artwork. Each Mylar film is scanned to create a bitmap image. The scanned bitmap could be directly converted to laser marking format with good image quality but the cycle time would be unnecessarily long with excessive marking line overlap.

For the best results, import scanned bitmap into AutoCAD as a positional template. Create a separate marking "logo" for each alphanumeric character and graphic image and, in AutoCAD, place each logo in position on a separate layer using the bitmap template as a positioning guide. A library of optimized logos facilitates creating programs from the scanned artwork, allows non-standard text kerning and line leading and assures low cycle time and high image quality. After all the logos are in place, the template layer is removed and the final CAD file is converted to the laser marking program format.

If the artwork already exists in a CAD file format, the image elements could be optimized using a separate library of logos. Every element including repetitive elements shared between drawings must be individually optimized. It will take considerably longer to convert large quantities of files if there is no guarantee that every element is optimized correctly. It is far more efficient to use the original AutoCAD file as the placement template for optimized logos.

Implementation of beam-steered laser marking requires a team effort. With cooperative implementation, manufacturing can assure product flow and integration with existing controls, the materials department assures plastics and coatings are appropriate for laser marking, and engineering will produce part marking programs with low cycle times and high quality images. Careful team planning, preparation and execution will result in a smooth application of laser marking technology and the associated benefits in manufacturing efficiencies, quality, and product value.

Lasermarkable Engineering Resins

B. M. Mulholland
Hoechst Technical Polymers, 8040 Dixie Hwy, Florence, KY 41042, USA

INTRODUCTION

Over a quarter century ago, lasers were thought of as "tools of destruction." From cutting through steel to use as weapons, lasers were powerful devices and thought to only be used that way. Of course today's lasers carry a much tamer connotation and a much broader usage basis that is ever growing. With the technological refinement that has occurred, lasers are now used in delicate surgical procedures, sight lines for alignment, precision grinding and cutting, lithography, communications and for the marking of products. Laser marking on plastics, that is the marking and decorating of plastic parts, is of particular interest.

Laser marking on plastics is growing in use. Bar codes and product lot data can currently be marked with lasers at high speeds on some commodity resins. However, of specific interest is the use of lasers to mark functional or decorative information on engineering resins. Many engineering resins because of their inert surface characteristic have been difficult in the past to mark *via* printing using ink. It is extremely difficult, for instance, to pad print on acetal without surface treating with very harsh chemicals. And even if the ink "adheres," the printed markings exhibit very poor wear characteristics and can be easily removed.

Laser marking is an excellent solution when problems in printing occur, or when there is a need for a truly indelible mark. For example in acetal, functional components such as cassette stereo buttons, hood and trunk release levers, or cruise control buttons can be laser marked with the functional description without fear of the identification rubbing off. In other applications, decorative marks can be made such as company logos and tradenames on acetal parts. These would include such items as car stereo trim plates, floppy disk shutters and other miscellaneous goods where the part supplier requires an indelible mark.

This paper focuses on the development of specialty grades of acetal copolymer that yield excellent sharp, clear images when laser marked. Specifically grades have been developed for laser marking on general purpose parts. Additional grades have been developed for laser marking on those applications requiring the utmost in ultra-violet (UV) light stabilization for both automotive and non-automotive applications.

TRADITIONAL MARKING & LABELING TECHNOLOGY

The most common methods for marking plastics today still include ink printing (both pad and ink-jet processes), ink filling, sublimation printing, embossing and stamping. Ink filling refers to the process of manually filling molded-in recessed areas with ink by injecting ink into these areas and wiping off the excess. Of course two-shot, two-color molding is another method to mark and label, albeit a very expensive one.

Of these traditional methods, ink printing is the most widely used. The primary benefits are the relatively low capital investment and the ability to print (pad print) on curved surfaces. The disadvantages to printing with ink are numerous and include:

- Non-permanent (relatively poor scratch and wear resistance and chemical resistance)
- Requires contacting the part surface
- Potential for smudged or illegible marks and labels
- Difficult to achieve on engineering resins
- Pre- and post-treatment processes typically required
- Environmental concerns including disposal of solvents and other chemicals
- Potential toxicity and/or flammability of certain solvents
- Maintenance of ink-jets and mechanical components
- Not flexible (manufacture of new die/transfer pad required for each new design)

ADVANTAGES OF LASER MARKING

In contrast to ink printing, laser marking of plastics provides excellent images without contacting the surface. Laser marking does require a higher capital investment. But economic analyses that take into account the facts of no consumable supplies required, no new dies/transfer pads required for design changes, speed of design change, and no hazardous waste generation for emissions or disposal, will generally favor laser marking depending on the number of components to be labeled. Even at somewhat higher per part costs, laser marking offers significant advantages that include:

- Indelible marks
- Non-contact to surface
- Extremely sharp images without smudging
- No pre- or post-treatments typically required
- No solvent use and no associated disposal
- Precision placement of marks and letters, even on irregular or curved surfaces

- Quick design changes *via* programmable software
- 2-D-symbology potential (ultra dense data capability)
- No adverse effect from part surface moisture
- Low operating cost (no consumable supplies to purchase such as ink)
- Low maintenance

LASER MARKING ON PLASTICS

The word laser is an acronym that stands for Light Amplification by Stimulated Emission of Radiation. The device itself emits a concentrated, precisely focused parallel beam of light. Lasers typically generate this light using an energy source, a lasing medium that allows the light to concentrate, and reflecting mirrors to direct the energy within the lasing medium. There are three types of lasers currently used to laser mark on plastics. They differ primarily in the wavelength of the resulting light energy. This is determined by the lasing medium used in the construction of the laser as described below.

TEA-CO₂ LASER

As the name implies, this laser uses carbon dioxide as the lasing medium (the acronym TEA stands for Transversal Excited Atmospheric pressure). The TEA-CO_2 laser operates at a relatively long wavelength of 10,600 nm. Images are typically produced using a mask that has the information etched into it. The laser fires its intense light through the mask. The resulting image is focused and redirected onto the object. The actual mark is achieved by the partial carbonization of the polymer due to the intense energy and creates an etch into the polymer with a depth typically in the range of 100 to 500 microns. The quality of mark is comparable to a dot matrix printer especially when marking at high speeds. TEA-CO_2 lasers are typically effective for simple coding such as lot numbering. However, high resolution graphics for appearance applications are better served by either of the other types of lasers. For acetal resins in particular, the major portion of the TEA-CO_2 laser energy is absorbed by the polymer matrix. This causes engraving of the surface without significant contrast.

N YAG LASER

In contrast to the carbon dioxide laser, the Nd:YAG laser uses a solid state medium of Neodymium Doped Yttrium Aluminum Garnet. The YAG laser, for short, can be operated either at 1064 nm or doubled frequency at 532 nm. The doubled frequency gives rise to sharper images. YAG lasers are typically interfaced with a computer to generate the graphics using a vector process achieved with focusing mirrors (Figure 1). The YAG laser in a sense writes on the surface of the plastic part. Since no masks are required, design change and flexibility are improved versus the TEA-CO_2 laser. And with the higher frequency, the distinctness of image is also far superior compared to the TEA-CO_2 laser.

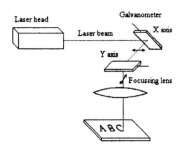

Figure 1. Nd:YAG laser schematic.

When operated at the 1064 nm wavelength, the YAG laser creates a mark by melting and foaming the polymer surface. Unlike the TEA-CO$_2$ laser, this surface interaction occurs only to a depth of about 50 microns. When excellent contrast is obtained (bright white mark on a black substrate), the foaming occurs to about 40 microns. By adjusting frequency and power, the amount of foaming can be altered and the color of the resulting mark can be made darker.

Frequency doubled Nd:YAG lasers operate with a wavelength in the visible region at 532 nm (green light) and typically affect pigments and other additives that absorb at that wavelength. The resulting color change is due to a photochemical process occurring to these pigments and additives rather than from melting and foaming of the polymer. However, if very high peak laser output is used, localized heating of the polymer can still occur resulting in melting and foaming.

YAG lasers are becoming increasingly popular for laser marking appearance applications. They are particularly suited for developing a light mark on a dark plastic part. To this end, lasermarkable acetal copolymer resins were specially formulated to enhance the contrast of a white mark on a black part using the YAG laser.

EXCIMER LASER

The Excimer laser generates UV light in the wavelength range of 193 to 351 nm. Here the laser marks totally by a photochemical process and the polymer matrix is not thermally loaded. Excimer lasers typically act on titanium dioxide or other mineral fillers to generate a dark mark on a white or light colored substrate. Relatively high levels of pigment or filler are necessary to achieve acceptable contrast. Since the process is photochemical, little to no etching occurs on the polymer surface. Marks penetrate to depths typically less than 40 microns. Excimer lasers have limited use for marking plastics today primarily due to being more expensive than Nd:YAG lasers and their limited ability to only produce a dark mark on a light substrate.

LASERMARKABLE ACETAL COPOLYMER RESIN

Because the Nd:YAG laser is the preferred marking device for developing high contrast marks on a dark substrate, development of a specialty lasermarkable grade of acetal copolymer was focused on that laser type. In particular, the objective was to develop a lasermarkable black formulation that yields the highest possible contrast when marked with the Nd:YAG la-

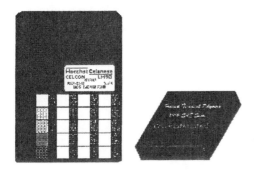

Figure 2. Lasermarkable acetal copolymer resins.

Figure 3. Marking with poor contrast.

ser. To that end, black lasermarkable acetal co-polymer was developed using patented technology. This resin yields extremely white, high contrasting marks as shown in Figure 2. Conventional black grades show little to no contrast as pictured in Figure 3. Typical applications for general purpose lasermarkable resins include appliance buttons and knobs, keypad keys, miscellaneous switches and incremented thumb wheels, and floppy disk shutters. An application such as functional and decorative markings on an electric razor take advantage of the wear and chemical resistance of the laser mark, as well as those same properties of the base acetal copolymer resin.

Building upon this patented technology, a UV stable, lasermarkable resin was developed for interior automotive and other applications. This resin combines the laser marking ability with the world-class ultra-violet light stability and can be laser marked with the Nd:YAG laser to produce excellent white marks with no yellowing caused by the UV stabilizer system. The mark produced on this UV grade is of the same high contrast as depicted in Figure 2.

UV stabilized, lasermarkable acetal copolymer meets all current automotive interior weathering requirements including the 1240.8 kJ/m^2 exposure requirement which is the highest standard in the industry. This resin is designed to be used in automotive interior functional components such as cassette stereo buttons, hood and trunk release levers, or cruise control buttons. In these applications, the parts can be laser marked with the functional description without fear of the identification rubbing off as currently can occur with ink printed components. In other applications, decorative marks can be made such as company logos and tradenames. An example is a car stereo trim plate marked with either the logo of the automaker or the stereo manufacturer.

THE FUTURE IN COLOR

While the initial focus of laser marking has been on developing a high contrast white mark on a black substrate, the possibility of developing a colored mark is intriguing, but challenging. A colored mark would no doubt expand the usage of laser marking and allow greater design flexibility for the customer.

Currently, the Excimer laser will yield a grayish to black mark on a light colored substrate. That is one option for color other than the white mark, albeit a limited one. The Nd:YAG laser offers seemingly more potential for marking colors. In acetal resin for example, laser marking with the YAG laser on a medium to dark color will yield a mark which is lighter in color and similar in hue. For instance, marking on a dark blue acetal part with the YAG laser will yield a light blue mark.

Building on this lasermarking technology, it may be possible in the future to expand the palette of colors when marking with the YAG laser. Possibilities include high contrast colored markings on a black substrate. Our initial successes in this area have included green marks or gold marks on a black substrate. What's more, it may be possible in the future to expand on this by developing technology which creates a colored mark on a colored substrate of different hue.

CONCLUSIONS

In conclusion, if your application calls for indelible, high resolution graphics, combining this new lasermarkable acetal copolymer resin with the Nd:YAG laser will produce the brightest, highest contrasting white marks on black molded parts, that can be achieved in industry today. Equally important, this combination truly eliminates any problems associated with ink printing adhering onto acetal and removes any worry concerning the mark wearing off.

REFERENCES

1   K. Kurz Haack, K. Witan, G. Lauck, "Non-Contact Marking by Laser Beam", Hoechst AG, translated from *Kunststoffe*, **83**, No. 7, p. 878-882 (1993).
2   S. Dass, Interaction of Lasers with Plastics and Other Materials, Society of Plastics Engineers, CAD RETEC '94 Proceedings, p. 236-244 (1994).
3   T. McKee, Laser Marking of Polyethylene and Other Polyolefins with Additives, Society of Plastics Engineers, CAD RETEC '94 Proceedings, p. 217-230 (1994).

The Enhancement of Laser Marking Plastic Polymers with Pearlescent Pigments

Jeff D. Babich
EM Industries, Inc., 5 Skyline Drive, Hawthorne, NY 10532, USA
Gerhard Edler
E. Merck, Frankfurter St. 250, Darmstadt, Germany

INTRODUCTION

The labeling or marking of manufactured plastic products is becoming increasingly important in today's marketplace. Products are labeled or marked for either identification or safety purposes. Such markings include expiration dates, bar codes, serial numbers, labels and company logos. Currently, printing (pad or ink-jet), embossing and stamping are the traditional techniques of affixing text, symbols and design elements onto the surface of plastic products. The new technology of laser marking provides an environmental and cost effective alternative to these traditional methods.

This paper will present information which is essential for the laser marking of plastics. The laser marking process and the advantages of operating a laser versus traditional methods will be discussed. The required equipment and pigment additives necessary to successfully mark plastics with lasers will be presented with samples demonstrating the effectiveness of each.

ADVANTAGES OF LASER MARKING

Currently, there are three types of lasers which are commonly used for the marking of plastic products:

Pulsed CO_2 laser (wavelength 10.6 μm)

Nd:YAG laser (wavelength 1064 nm or 532 nm in the case of frequency doubling)

Excimer laser (UV laser, wavelength 193 nm - 351 nm)

The laser generates a high speed mark, with sharply defined characters which are possible on smooth, soft or irregular surfaces. With a pulse CO_2 mask laser up to 5000 markings per minute are possible, even on moving parts. A Nd:YAG laser can produce a mark up to 2000 mm per second. Laser marking is a highly flexible process, in terms of character programmability with the Nd:YAG laser and automatic mask change systems used with a CO_2 laser. Because laser marking is a non-contact process with no mechanical wear, service and running costs are minimal.

Laser marking is an ink-free process which produces a permanent rub-fast, solvent resistant, scratch proof mark with no waste, and is free of flammable or volatile solvents and their subsequent disposal. These permanent markings cannot be removed without damaging the product, which is ideal for safety concerns. The laser marking process does not require any form of surface treatment, in fact, a film of water on the product will have no effect on the marking process. Since the mark is etched into the plastic product, there is no problem of adhesion, even on polyethylene and polypropylene.

In laser marking plastics, the quality of a laser mark can only be achieved with very few plastic resins in their virgin state (e.g., PVC). The majority of resins used in packaging, including polyolefins and styrenics, have up to now proved difficult or impossible to mark adequately using laser technology. Plastics with a low absorption level of laser light show practically no reaction to laser bombardment. The absorption of the plastic material is too low for the laser beam, thus the beam passes through the plastic material without creating a visible mark. Only after the laser intensity has been increased drastically or the process repeated several times is it possible to produce a faint, very low contrast mark. However, those procedures can generate stress on the resin so great that polymer degradation occurs. The incorporation of pearlescent pigments to these low absorption plastics allows the resin to become receptive to laser light, consequently a high contrast, visible mark is achievable at a relatively low laser intensity.

WHAT IS PEARLESCENCE

The use of pearlescent pigments vastly increases the possibilities of color design. In comparison to other pigments, organic or inorganic, pearl pigments possess lustre as an additional property. The pigments consist of a flat mica platelet, which is coated with one or more very thin metal oxide layers. In all the pearlescent pigment series (Silver White, Interference, Gold and Earthtone), effects ranging from satin to glitter can be obtained by modifying the particle sizes of the mica. Due to the platelet form and the higher refractive index of the metal oxides in comparison with the surrounding medium, a pearl luster effect is obtained. Pearlescent pigments are often used without other added colors, but when combined with various transparent pigments, the possibilities of coloring plastics are practically limitless. Pearl pigments can be used with all thermoplastic applications, and are appropriate for some thermoset uses.

LASER MARKING WITH PEARLESCENT PIGMENTS

Although pearlescent pigments are commonly known for their decorative enhancement to plastic products, the pigments can now be used in a functional capacity. Today, due to further development of pearl lustre pigments and close cooperation with the manufacturers of laser marking systems, it has become possible to laser mark several types of plastic materials while retaining the flexibility in color design. The quality of the mark depends upon the plastic resin involved, the type of laser instrument and the pearlescent pigment used as well as its concentration. Recommended concentrations of 0.2% to 1.0% pearlescent pigments are adequate for the laser marking of plastics. Increasing the pigment concentration results in a darker mark, while the penetration of the laser beam into the plastic is reduced. The optimum pigment concentration for a well contrasted, sharply defined mark depends upon the type of plastic and thickness of the plastic product. These variables determine whether a black, white or grey contrasting mark will be produced.

When laser marking a product is a requirement, but the pearlescent appearance is not necessary or desired, the consumer now has these options.

(1) Add a pearlescent pigment to the product, in very small percentages (0.2% to 0.5%), with opaque pigments, for example TiO_2, to cover up the pearl effect. This has no negative effect on the laser marking result.

(2) Incorporate Lazer Flair™ a new pigment introduced by EM Industries which enhances laser markability without the pearlescence appearance, into the product at a recommended loadings of 0.2%- 1.0%. Even with the addition of 1.0% Lazer Flair™ into the plastic compound, the color appearance of the product remains basically unchanged.

The phenomenon of laser marking plastic materials, which were otherwise unreceptive to a laser beam, with the assistance of pearlescent pigments is yet to be entirely explained. Theories have been developed as to what function the mica-based pearlescent pigments have in allowing a once non-markable resin to become receptive to the laser beam. One theory is that the laser produces a color change by partial carbonization of the plastic around the pigment particles. The carbonization only develops around the pigment particles near the surface so the overall stability of the plastic remains unaffected. In many cases a darkening of the pearlescent pigment also occurs. Carbonization and/or bubble formation on the surface of the plastic product may also result in the production of a grey mark.

CONCLUSIONS

Increasing requirements in the plastic industry regarding the quality of text or symbols, the speed of the process, environmental concerns, as well as flexibility and the need to reduce production costs have influenced the rapid development of the laser marking process. Laser marking of certain plastics in the past was found difficult due to the inability of the resin to react with the laser beam. Studies have found that the addition of pearlescent pigments into the non-receptive resin systems transforms them into laser sensitive resins capable of producing a visible mark. Markings in plastics produced with a laser aided by pearlescent pigments are rapid, irreversible and sharply defined.

ACKNOWLEDGMENTS

The authors would like to acknowledge with appreciation the contributions of research by Dr. Manfred Kieser and Reiner Delp of E. Merck, Darmstadt Germany. Recognition is also extended to Dr. Terry McKee of Lumonics, Inc. and Dr. Shiv Das with Lasertechnics for laser marked samples used in this study.

REFERENCES

1   Roman Maisch, Manfred Weigand. Pearl Lustre Pigments. Darmstadt, Germany, Verlag Maderne Industris AG & Co., 1991.
2   Manfred Weigand, Laser Marking with Pearl Lustre Pigments, E. Merck Darmstadt, Germany.
3   Gehard Edler, Iriodin/Afflair™ for Laser Marking, E. Merck Darmstadt, Germany.
4   Brian Norris, Laser Coding Makes Its Mark, *Package Technology and Engineering*, Feb 1994, 33.
5   John Myers, Lasers Make Their Mark on a Variety of Plastic Parts, *Modern Plastics*, Oct 1993, 24-26.

Index